D0216231

WORLD DISASTERS

FITZROY DEARBORN PUBLISHERS
LONDON • CHICAGO

CHABOT COLLEGE LIBRARY

WORLD DISASTERS

TRAGEDIES IN THE MODERN AGE

KEITH EASTLAKE

HENRY RUSSELL

MIKE SHARPE

Ref
D
24
.E19
2001

© Brown Partworks Limited 2001

All rights reserved, including the right of
any reproduction in whole or in part in any form

For information, write to

Fitzroy Dearborn Publishers
310 Regent Street
London W1B 3AX
UK

or

Fitzroy Dearborn Publishers
919 North Michigan Avenue
Chicago, Illinois 60611
USA

British Library Cataloguing in Publication Data.
A catalogue record for this book is available
from the British Library

A Cataloging-in-Publication record for this book
is available from the Library of Congress

ISBN 1-57958-318-0

This edition first published by
Fitzroy Dearborn Publishers 2001

Printed and bound in Spain by Artes Gráficas Toledo S.A.U.
D.L. TO: 608-2001

While every effort has been made to trace the copyright of the photographs and
illustrations used in this publication, there may be an instance where an error has
been made in the picture credits. If this is the case, we apologize for the mistake
and ask the copyright holder to contact the publisher so that it can be rectified.

Page 1: *French emergency staff remove passengers' belongings from the wreckage of the two trains involved in the crash at Argenton-sur-Creuse, France, on August 31, 1985.*

Pages 2-3: *US Coast Guards and FBI investigators watch as a tangled piece of the wreckage of the TWA Boeing 747 is raised from the waters off Long Island, New York. The 747 exploded in mid-air, killing all on board.*

Below right: *The roll-on, roll-off car ferry* Herald of Free Enterprise *lies capsized just outside Zeebrugge harbor. The ferry overturned when sea-water entered through the open main bow door. Nearly 200 people died in the disaster.*

For Brown Partworks Limited
Editors: Shona Grimbly, Ian Westwell
Design: Brown Partworks, Wilson Design Associates
Picture research: Wendy Verren, Susannah Jayes
Production: Matt Weyland

CONTENTS

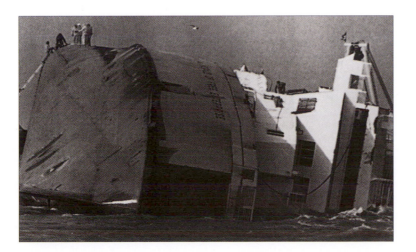

INTRODUCTION

Below: The liner Morro Castle was destroyed by fire on September 8, 1934, off the coast of New Jersey. Here, sightseers flock to see the burned-out vessel beached at Asbury Park, New Jersey.

Disasters – if we knew what caused them, we could prevent them from happening, or at least reduce the number that occur. But we don't know, at least not without the benefit of hindsight, and this rarely helps us make plans for the future because these events are so versatile and variable – one is never quite like any other.

Disasters resist categorization – they are by their nature unpredictable and disorderly. What they have in common – apart, of course, from the havoc and often dreadful human suffering they leave in their wake – is that they have all been sparked by a combination of often unforeseen circumstances. A single ingredient can usually be dealt with before it gets out of hand and is seldom sufficient to cause a catastrophe, but two or more ingredients combining are lethal, particularly when they occur suddenly or create a situation that has never been encountered before.

There are many contributory factors that can lead to a catastrophe. The most common is outright human error. We all make mistakes. Some of them can be put

Right: An injured passenger is removed from the wreckage of a Dutch train involved in a collision at Schiedam, May 1976.

Below: French rescue workers and crash investigators mill around the two trains involved in the accident at Argenton-sur-Creuse, August 1985.

right straight away and no one notices the difference; others we can cover up. Some may land us in trouble – they may even cost us our jobs. A few end in the deaths of one or more people: a train driver fails to observe a stop signal (as at Lewisham, London, in 1957) or a warning to reduce his speed (Argenton-sur-Creuse, France, 1985); one pilot retracts the leading wingflaps before his plane has attained sufficient airspeed, causing it to stall and crash (Staines, England, 1972), another makes a spur-of-the-moment decision to take a scenic detour which ends in disaster (Mount Fujiama, Japan, 1966); soldiers mistake the sound of an accident in a road tunnel for the noise of enemy artillery and then seal the entrances, so condemning up to 2700 of their comrades and civilians to death by burning or sudden asphyxiation (Afghanistan, 1982).

However, mistakes or errors of judgment alone are not usually enough to kill so many people – this book describes the deaths of nearly a million individuals, most over the last 100 years – nor destroy such vast amounts of property. There is nearly always in addition at least one of four other factors. The first of these are sometimes known as "Acts of God." These are natural forces that either cause disasters themselves – for example, the iceberg that collided with the RMS *Titanic* in 1912 and the thunderstorm that caused an Eastern Air Lines Boeing 727 to crash at New York's Kennedy Airport in 1975 – or take the role of catalysts, turning small mishaps into major destructive forces, as when sparks generated by lightning fanned by strong winds burst into rampaging forest fires such as those that swept through several states of the United States during the summer of 2000.

Another frequent contributory factor is negligence or bureaucratic stupidity. Thousands have been killed by fires in cinemas, night clubs, and other commercial

Right: The fire at Bradford City football stadium in England. A total of 56 people were burned to death when a timber stand caught fire on May 11, 1985.

Below: In Philadelphia, USA, fire swept through a residential district when police bombed the headquarters of Move, an "anti-society" radical organization, on May 15, 1985.

premises because the doors have been barricaded to prevent people from getting in or out without paying. One of the most harrowing accounts in this book is that of the 1985 fire at Bradford City football ground in northern England. Here, 12 of the 56 victims were crushed to death as they tried to crawl out under turnstiles which had been locked when the gatemen went off duty to stop latecomers getting in for free. Sadly, this is not a disgraceful exception but a recurrent theme in the history of such tragedies. It seems that in many cases the regard for public safety is surrendered to financial necessity or a simple lack of awareness of potential problems.

Equally, those placed in positions of responsibility can occasionally be defeated by technology. When two trains traveling in opposite directions moved onto a single track near Aamodt, Norway, in January 2000, signalmen in the network control room could see the danger on their computer screens but could not find the right mobile phone numbers to contact the drivers and warn them in time to prevent a head-on collision that left more than 30 people dead.

The fourth element, and in many ways the most shocking, is malice, which can include acts of war. This most frequently takes the form of arson or sabotage, sometimes both together and often because of an individual grudge, as in the case of the Italian cruise ship *Achille Lauro*, which was hijacked by Palestinians

Right: *Rescue workers inspect the scene of destruction left when an El Al jumbo jet crashed into an apartment block in Amsterdam on October 4, 1992.*

Below: *Japanese rescue personnel search through the wreckage of the Boeing 707 that crashed into a forest at the foot of Mount Fuji on March 5, 1966.*

off the Horn of Africa in 1985 and eventually destroyed by fire. Some of the worst cases in this book are the work of psychotics who were getting their own back for some real or imagined slight. One madman blew up an after-hours drinking club in central London in 1980, killing 37 people, because he thought that he had been overcharged for a rum and Coke. Another who had been thrown out of a hostel in Queensland, Australia, in July 2000 went back later to burn the place down, killing 18 backpackers in the process.

Some saboteurs conspire to destroy land and buildings for business reasons. The bush fires that burned throughout Indonesia for the greater part of 1997 were started deliberately by owners of forests and plantations who wanted to clear their land so that they could use it for more lucrative crops. The razing of La Fenice in Venice in January 1996 is a most unusual case – the arsonists are thought to have set fire to the historic opera house in order to prevent the owners from being able to invoke a penalty clause for late completion of a building works contract. This fire was an organizational triumph – there were no injuries or loss of life because it was planned meticulously to do its work while there was no one in or near the building; the blaze did the greatest amount of damage in the shortest time and the perpetrators, allegedly linked to organized crime gangs, were never brought to justice by the authorities.

Far right: The Italian liner Achille Lauro *on fire and adrift in the sea off the Somali coast the day after 1000 passengers and crew were rescued from the ship's lifeboats in November 1994.*

Below: The 1957 disaster at St John's, Lewisham, was caused when the crew of an express train failed to notice signal changes.

The quintessence is bad luck. This includes the disastrous failure of working parts at the worst possible moments, as for example when metal fatigue caused an engine to drop off an El Al Boeing 747 just as it was taking off from Amsterdam's Schipol airport in 1992. When a single piece of debris was left on a runway at Paris Charles de Gaulle in July 2000, the odds against an aircraft's tire passing directly over it during take-off must have been astronomical. Even when it did, it was still unlikely that the impact would explode the wheel, puncture the aircraft's engine and fuel bays, and thus start a fierce blaze, causing the Air France Concorde to crash into a hotel near the airport. Yet, this is what did happen, with the loss of all of the aircraft's passengers and crew – and several individuals on the ground.

In November 1986 there was a small fire in a chemical factory on the banks of the Rhine in Basel, Switzerland. The fire brigade put it out; no one was hurt; the incident seemed a routine matter. A few days later it emerged that the water used to douse the flames had been contaminated with insecticides before draining into the river. This caused an immense ecological disaster which blighted the Rhine downstream far beyond Cologne. Here it seems there was no real culprit, just a malign fate.

Even those who intended to cause damage have done no more than start their fires – they have seldom had any clear preconception of the extent of the destruction that would eventually be caused by their deeds. Chance has always played a major part. US President Franklin D. Roosevelt and his chiefs of staff, who masterminded the 1945 Tokyo firebombing massacre, could not have foreseen that their actions would lead to the deaths of 200,000 people.

In one of the most extraordinary cases, that of the destruction of the steamship *Morro Castle* in 1934, all five elements seem to have conspired. The disaster was caused principally by a fire which broke out in a locker below decks. This imminent danger might have been brought under control quite quickly, but the vessel's smoke detectors had been turned off and the

nearby fire extinguishers were sealed up, probably due to bureaucratic incompetence. More lives might have been saved if the vessel's officers had not refused to raise the alarm for fear of worrying passengers. Just before the fire, the captain died of a heart attack, an unfortunate piece of extremely bad luck, and the chief officer who took over his duties lacked the necessary experience to deal with the mounting crisis – this could charitably be described as human error, although the man in question, possibly a scapegoat, later served a prison sentence for negligence during the incident. As the flames took hold, most of the crew were busy battling through a wild storm on a return trip to New York from Havana, Cuba. The 40mph (64km/h) wind swept flames through the whole superstructure of the *Morro Castle*, arguably an "Act of God." It was later suggested during the investigation into the sinking that the ship's chief radio operator was probably an arsonist, suggesting that the vessel's sinking involved a measure of personal malice.

Another unusual aspect of the *Morro Castle* disaster was that some good came out of it – much tighter regulations were brought in to govern the safety of US passenger ships. More often, however, the lessons are

Below: An aerial view of the rescue vehicles and wreckage of the Eastern Airlines Boeing 727 that crashed at JFK airport, New York, on June 24, 1975.

not learned at all or else they are quickly forgotten. The Dutch rail disaster in January 1962, when two trains collided in fog at Harmelen killing 93 people, made it clear to the authorities that an automatic train protection (ATP) system was urgently needed. However, it had been installed on only about a quarter of Dutch track by May 1976 when 24 passengers were killed in almost identical circumstances on two trains that crashed at Schiedam. Even after this, 10 years later only 60 percent of Dutch railroads had ATP installed.

Some disasters remain a mystery. One such was the 1985 fire in Philadelphia, in which 11 people were killed after police tried to clear a building that contained the headquarters of a radical anti-establishment group called Move. No one ever found out whether the explosion that caused this tragedy was an accident or if the police had taken a deliberate decision to use extreme force, no matter what the consequences to property and human life.

All over the world and throughout history, disasters have been a fairly common way of violent death. Those authorities responsible for the safety of the public have a legal duty to prevent such events. However, they do not have bottomless purses. Arguably, it would be possible to reduce the threat of disasters to a wholly insignificant probability – given time and huge amounts of money. Nevertheless, no amount of forward planning or finances can entirely remove the danger, particularly when many incidents are the product of the vagaries of the human mind and the weather. Technology, then, is not a panacea, and disasters will always occur despite humankind's best efforts. Since it seems we cannot eradicate them, all we can do is hope that we are not in the firing line when they do strike.

Above: A Malaysian fireman uses a tree branch to fight the forest fires that rampaged throughout Southeast Asia during the autumn of 1997.

Right: On the night of January 30, 1996, fire devastated Venice's famous opera house, La Fenice.

The devastation brought about by catastrophes, whether they are the product of natural forces, the work of a group or individual, or random and unforeseen events, is not confined to the modern world. For as long as humankind has inhabited the planet, we have fallen victim to such incidents. Yet even in the modern world ordinary people and those who represent them have continued to view catastrophes as rare or in some way often unavoidable. This was never wholly true before the modern age and is not so today.

As the following incidents show, catastrophes can have many causes, some deliberate and others inadvertent. For example, the Great Fire of London in 1666 was brought about by a carelessly tended fire in one of the capital's small bakeries. The blaze, coupled with closely packed wooden buildings and limited firefighting facilities, led to the destruction of large swathes of London. Equally, for many years the fires that razed much of Rome in AD 64 were thought to have been deliberately instigated on the orders of Nero, the emperor of the time.

Whatever the reasons behind a catastrophe, all have several things in common. They bring death, destruction, and misery. However, they also lead to a search for the cause or causes that provoked the incident and, more often than not, attempt to apportion blame. Whatever the age of a catastrophe, these issues have remained constant.

Background: The shattered remains of the USS Maine *lie on the bed of Havana harbor in Cuba following the mysterious explosion that wrecked the warship in February 1898.*

WARNINGS
FROM THE PAST

ROME

JULY 18, AD 64

Left: Emperor Nero watching Rome burn. At the time Nero was generally believed to have started the fire himself. Suetonius, the contemporary historian, reported that Nero "set the city on fire so openly that his attendants were caught with burning torches in their hands."

The Great Fire of Rome is one of the earliest fire disasters in history about which there is reliable documentary evidence. Although details are scarce – neither the number of deaths nor the full extent of the damage to property has ever been accurately determined – the fire is known to have been a wide-ranging and cataclysmic inferno with important consequences for the city and its people.

Fire had long been one of the greatest hazards in the tightly-packed wooden houses of ancient Rome, and had caused extensive damage at various times in its history. The first Roman Emperor, Augustus Caesar, who ruled from 27 BC to AD 14, made fire prevention one of his top priorities. In 21 BC, he organized the first Roman fire brigade, a band of irregulars comprised of slaves under the command of officials who until then had been in charge of streets and markets. This did not solve the problem, however, and after a bad fire in AD 6 Augustus established the first corps of professional firemen – known in Latin as *vigiles* – made up of seven squads or cohorts of 1000 freemen each.

The Great Fire broke out on the bright moonlit night of July 18, AD 64, and raged unchecked for over a week, totally destroying three of Rome's 14 regions and partially destroying seven others. Then, as now, it was commonly believed that the fire had been started by the Emperor Nero (who ruled from AD 54 to 68). He was away from Rome at the time, and his alibi seemed suspiciously convenient.

Alarmed that the population seemed to be blaming him, Nero tried to blame the fire on the Jews. Although the Emperor's second wife, Poppaea Sabina, maintained that they were innocent, the Jews were widely persecuted in the aftermath of the blaze. The most famous victims of the ensuing violence were Saints Peter and Paul. Although by this time the Roman people were starting to differentiate between Jews and Christians, these two were not different enough to escape martyrdom. The climate of hatred and retribution created by the Fire continued into AD 65 – Nero remained unpopular, and when a plot to assassinate him was uncovered, the poet Lucan and numerous senators were executed.

Above: After the Great Fire that destroyed most of Rome Nero punished those he held to be responsible.

Above: Nero amid the ruins of Rome. Nero took measures to control the fire and to find shelter for the homeless.

However, no historian now believes that Nero was responsible in any way for the fire. Nor, indeed – contrary to popular legend – did the Emperor fiddle while Rome burned. The truth of the matter seems to have been that, on hearing the news from his capital, Nero rushed back and took strenuous measures to bring the fire under control and to supervise the provision of shelter for the homeless. Afterwards, Nero's initiatives to rebuild the city with broad thoroughfares acting as firebreaks contributed in no small measure to the modern-day appearance of Rome. Nevertheless, there were several other major fires in Rome, most notably in AD 80 during the reign of the Emperor Titus, when the Capitol, the Pantheon, and Agrippa's Baths were among the buildings destroyed.

THE GREAT FIRE OF LONDON

SEPTEMBER 2–5, 1666

Below: People took to the river in an attempt to escape the flames.

Bottom: The fire as seen from the south bank of the River Thames.

The Great Fire of London was the worst fire in the city's history. It destroyed a large part of the City of London, including most of the civic buildings, old St Paul's Cathedral, 87 parish churches and about 13,000 houses. At least nine people were burned to death. Some important historical insights into this appalling disaster are contained in the diaries of Samuel Pepys (1633–1703), who was living in London at the time.

The fire broke out accidentally at King Charles II's baker in Pudding Lane near London Bridge when a spark from the ovens ignited the hay in a neighboring inn yard. The blaze spread quickly, partly because of the density of the surrounding wooden buildings, and partly because so many of them were full of combustible matter such as pitch and tar, oil and brandy. By the following morning, half a mile of the River Thames waterfront was ablaze.

It was the end of summer and the weather had been hot and dry for some time. According to Pepys, this made "everything, after so long a drought . . . combustible, even the very stones of churches." A strong east wind fanned the flames, which raged during the rest of Monday and part of Tuesday.

At first, people tried to save their possessions as well as their lives. Some loaded up horses and carts full of their belongings and set off to the fields of Hampstead and Highgate; others moved their belongings from one burning house to another still unaffected nearby. But their efforts were futile: the fire was rampant and took all before it. Soon the Thames was swarming with barges full of people who by now were taking with them only what they were able to carry. Even on the water, however, they were not entirely safe: the conflagration was now so strong that, as Pepys describes it, "with one's face in the wind, you were almost burned with showers of firedrops."

Pepys and others advised the King to blow up the houses in the fire's path to create a firebreak, but the conflagration outran them. Sir Thomas Bloodworth,

Southwark

Lord Mayor of London, lamented: "What can I do? I have been pulling down houses. But the fire overtakes us faster than we can do it."

On Wednesday the fire slackened and began to burn itself out; on Thursday it was largely extinguished, but in the evening of that day the flames again burst forth in the Temple, one of the Inns of Court. Some houses in Tower Street, next to the Tower of London, were at once blown up by gunpowder and after that it became easy to quench what fire remained. The fire was finally mastered.

Within a few days of the fire, King Charles II was presented with three different proposals for the reconstruction of the city – two were drawn up by the architects Christopher Wren and John Evelyn, and the other by the city's Surveyor of Works, Robert Hooke. Although they all contained plans to regularize the streets, none was adopted and nearly all the old thoroughfares were retained. Nevertheless, Wren's greatest achievement, the new St Paul's Cathedral, did get built and still stands today.

Below: The devastation caused by the fire in the city of timber-framed houses was horrifying. When the city was rebuilt, less combustible materials were used, such as brick and stone.

Above: Fanned by a strong east wind the flames spread rapidly through the close-packed streets of wooden houses. The terrified Londoners were able to snatch up only a few possessions as they fled in fear of their lives.

LA COMPANIA CHURCH, SANTIAGO, CHILE

DECEMBER 8, 1863

Below: The city of Santiago, capital of Chile, with the Andes mountains behind it. Here, one of the worst ever fire disasters happened in the Church of La Compania.

In the world's worst fire disaster in a single building, approximately 2500 people – nearly all women and children – were burned to death as flames spread quickly through La Compania church in Santiago, the capital of Chile. That there were very few men among the victims is explained by the fact that the tragedy occurred on the Feast of the Immaculate Conception. This day – one of the most important dates in the Roman Catholic calendar – is of particular significance to women because it celebrates the doctrine that Mary, the mother of Jesus, was preserved from original sin.

The interior of the 17th-century La Compania church – famously beautiful in itself – had been specially decorated for the occasion with colored lamps which had been hung in groups in every nook and cranny of the building. The parishioners were keen to celebrate the festival as memorably as they could, while the parish priests are thought to have been annoyed by remarks attributed to a Papal Nuncio, who compared the Chileans' religious celebrations unfavorably with those in Rome. One local priest, Father Ugarto, was particularly annoyed by adverse criticisms of the lighting in Chilean churches: he is reputed to have

Above: The ruins of the 17th-century Church of La Compania after the fire. The church could hold 3000 people, and it was full to capacity on the Feast of the Immaculate Conception when the fire broke out.

responded with the words: "I will give the Nuncio such an illumination as the world has never seen."

As a consequence, on December 8, the day of the Feast, there were some 20,000 of these candle-powered lamps spread all around the interior of La Compania. Priests and lay helpers had spent several hours during the afternoon lighting them all in readiness for the evening service.

The service was due to begin at 1900 hours. At a couple of minutes before the hour, the church was full to capacity, with about 3000 people squatting on prayer mats – there were no chairs or pews. On the altar, the centerpiece of the service, was a statue of the Virgin Mary with her feet set upon a crescent moon made of canvas and wood: the moon flickered atmospherically from the light of the lamps inside it.

On the stroke of the hour, the moon at the feet of the Virgin Mary caught fire and flames spread quickly through the building. Only the main door was open and this was not large enough to let all the congregation out in time. The door of the other potential escape route, through the sacristy, had been shut by the priests behind them as they fled to safety. Chaos reigned as the congregation stampeded.

The church was completely gutted in less than 30 minutes. Of the victims, many were under 20 years old. The bereaved men of the city blamed the priests who they thought had made good their own escape and left the congregation to die.

THE GREAT CHICAGO FIRE, USA

OCTOBER 8–10, 1871

Four square miles of Chicago, including the whole of the business district, were destroyed in October 1871 by a blaze that raged across the city for nearly three days. The conditions were ideal for a fire – there had been a long, hot summer and most of the buildings were built of wood. Starting in a southwestern suburb, flames spread rapidly northeastward, leaping the Chicago River and dying out only when they reached Lake Michigan. About 250 people died, 90,000 more were made homeless, 18,000 buildings were destroyed and almost 200 million dollars' worth of property was lost.

No one really knows how the Great Chicago Fire began, but according to popular legend it broke out in DeKoven Street, southwest Chicago, in a barn belonging to a woman called Mrs O'Leary. According to the story – which was always strenuously denied by the woman herself – at about 2100 hours on the night of October 8, her cow kicked over a lighted kerosene lamp which set fire to a pile of hay and wood shavings.

From there, the flames spread very quickly, fanned by a strong wind blowing from the west. The breeze pushed the fire away from the O'Learys' house – which miraculously emerged unscathed from the ensuing

Left: The Chicago fire was popularly supposed to have been started in a barn, when a cow kicked over a kerosene lamp which set fire to a pile of hay.

conflagration – and off in the direction of the downtown area. By midnight, the fire had reached the Chicago River. This should have been a natural barrier and the end of the matter, but the wind ferried blazing pieces of wood across the river and, once on the other

Left: Even the Chicago River could not stop the progress of the fire. The wind blew pieces of burning wood across the river to the other side where there was an abundance of combustible materials.

Right: The Chicago firemen fought the flames tirelessly, but with their limited resources they could not put out the conflagration.

side, the fire continued its progress toward the heart of the Midwest capital. Downtown there was an abundance of fuel – grain in great silos, lumber yards full of wood, coal depots and liquor distilleries. The fire services proved completely inadequate. There were only 17 horse-drawn fire engines and 23 hose carts, nowhere near enough to put out a fire of this size.

The *Chicago Daily Tribune* described how the flames would "hurl themselves bodily several hundred feet and kindle new buildings. The whole air was filled with glowing cinders, looking like an illuminated snowstorm. Fantastic fires of red, blue, and green played along the cornices of buildings."

The First National Bank building was popularly supposed to be fireproof and, although it did not catch light, it became so hot that its iron framework expanded until its external walls collapsed. In the downtown area, every hotel, every theater, and every bank was destroyed, together with the Grand Opera House and the City Hall. The fire only stopped when it reached the shores of Lake Michigan.

Despite the efforts of the citizens, the fire was never brought under control but simply exhausted itself. As the smoke cleared, the exhausted firefighters were amazed to see one house – the home of Mahlon Ogden – still largely intact amid a sea of ashes. The story is that, on seeing the flames approaching, the family soaked all their carpets, blankets, and sheets in water and hung them over the walls and roof, thus preventing their home from catching light.

PRINCESS ALICE, THAMES

SEPTEMBER 3, 1878

As the wreck of the *Princess Alice* shows, shipping disasters are not confined to the high seas. The *Princess Alice*, launched in 1865, was navigating the placid Thames River in London when catastrophe overtook the 251-ton vessel. Those on board were making their way to Gravesend to enjoy one of the last days of summer. For many that September day was destined to be the last day of their lives.

The *Princess Alice*, a vessel belonging to the London Steamship Company, was involved in a collision with a collier, the *Bywell Castle*. The collision and the resulting loss of 640 lives shocked the nation. One report states that it was one of the most fearful disasters of modern times – a scene which has no parallel on the river. The report went on: "the river for 100 yards (300m) was full of drowning people screaming in anguish and praying for help." Many of those on board the *Princess Alice* were families and there were many young children lost in the disaster.

The *Princess Alice* had only just left its moorings in London and had reached a point about a mile below Greenwich. Those on board could have had no warning of what happened next. The 1376-ton collier sliced right into the river steamship, which was badly holed and sank rapidly.

The ship's master, Captain Grinstead, was one of those who did not survive. When a final head count was made, there were 200 survivors.

Below: When the collier the Bywell Castle *sliced into the pleasure steamer the* Princess Alice, *many people were killed instantly. Others drowned in the river before any rescuers could arrive.*

HMS VICTORIA, MEDITERRANEAN

JUNE 22, 1893

The loss of HMS *Victoria* in the eastern Mediterranean in 1893 remains the worst peacetime disaster that the British Royal Navy has ever suffered. The 10,740-ton *Victoria* was launched in 1887 and was something of an experimental ship. At the time of the sinking the warship was armed with two guns each weighing 111 tons. These were so heavy that their barrels tended to droop down under their own weight and it was believed unwise to fire them with their full charges of powder.

The loss of the *Victoria* was in large part due to the misjudgment of one man, Vice-Admiral Sir George Tyron. Tyron, who was commander-in-chief of the Royal Navy's Mediterranean Fleet at the time of the sinking, was an experienced officer. He had commanded the navy's first ironclad warship, the *Warrior*, and was considered a naval warfare expert.

The *Victoria* was Tyron's flagship on the day of its sinking. Tyron had split the fleet in two and was carrying out maneuvers. The *Victoria* and the other vessel involved in the incident, HMS *Camperdown*, were each leading their respective halves of the fleet. Tyron ordered the two lines of ships to turn towards each other. The captains of both leading vessels complied with his order, yet it was plain to many watching that the two ships were far too close to each other for them to complete the change in direction without colliding.

The two ships did collide and the *Victoria* sank with the loss of 359 men, including Tyron himself. Fortunately the other vessels of the fleet were on hand to rescue the 284 survivors.

Below: HMS Victoria, *the flagship of the British Mediterranean Fleet, sinking after it collided with the vessel* Camperdown *while carrying out maneuvers off Tripoli.*

USS MAINE, HAVANA

FEBRUARY 15, 1898

No one knows exactly what caused the explosion that led to the sinking of the 6682-ton battleship USS *Maine*, but there have been a number of theories, some more probable than others. What is certain is that the sinking of the warship on February 15, 1898, resulted in the deaths of 258 enlisted men and three officers. It was the worst loss of life suffered in peacetime by the US Navy up to that date, and it led to the United States declaring war on Spain in April 1898.

Far right: The USS Maine, *which was anchored in Havana harbor, Cuba, when it was destroyed by an explosion in February 1898. The incident led to the United States declaring war on Spain.*

Below: The wreck of the USS Maine *after it had been destroyed by the explosion.*

The *Maine* was anchored in Havana harbor, Cuba (which was a Spanish colony at the time) to protect US citizens on that increasingly volatile island and to evacuate them if the fighting between the Cuban insurgents and the Spanish authorities took a turn for the worse. The United States had expressed support for the insurgents, but there were no hostilities between the United States and Spain when the *Maine* sailed into Havana on January 25.

Both sides observed the usual diplomatic formalities, the US ship exchanging salutes with the Spanish warships at anchor. The *Maine*'s commander, Captain Charles Sigsbee, was fully aware of the sensitivity of his mission and took measures to ensure the safety of his ship – sentries were posted and steam was maintained in two rather than just one of the ship's boilers in case a quick getaway had to be made. Ammunition for some of the battleship's secondary armaments was kept ready and all visitors were carefully watched.

Shortly after Sigsbee retired to his cabin at 2130 hours on the evening of February 15, the *Maine* was rocked by a violent explosion – powerful enough to smash the ship's powerplant and break windows in Havana town. Other officers then reported a second explosion a few moments after the first. Sigsbee decided to flood the ship's magazines to prevent further explosions, but the crew told him that water was pouring into the ship and the magazines were already flooding. Fires were also raging in a mess hall amidship. When he heard that the forward magazine was threatened by fire, Sigsbee gave the order to abandon ship. At the roll call taken later, it was found that there were only 94 survivors, 55 of them injured.

Investigators initially believed that a magazine had exploded by accident, but senior US government officials were looking for an excuse to declare war, and wanted to use the incident as a pretext to do so. After hearing the reports of the divers who had investigated the wreck, the official inquiry decided that an underwater mine had been responsible for the catastrophe. The divers reported that hull plates had been blown inward, suggesting an external source of the explosion. These findings were confirmed in 1911 when the *Maine* was raised.

The sinking of the *Maine* galvanized public opinion against Spain in the United States, and the rallying call "Remember Maine – to hell with Spain" was heard on all sides. On April 29, 1898, the US Congress declared war. The Spanish–American War, that was to end in a US victory, had begun.

By the beginning of the twentieth century humankind was moving into an era dominated by rapid technological progress. Many saw technology as the panacea to all of the world's ills. Its application would dramatically reduce disease, illness, and poverty, generate previously unimaginable wealth, and vastly improve the quality of life of those able to take advantage of it. People would live longer, enjoy greater happiness, and be in some way protected from harm. Many of these claims proved true to a large degree, at least to those who had access to the new technology. However, other events rapidly proved such wide-ranging claims to be wildly over-optimistic, and various catastrophes quickly undermined any unquestioning faith in the infallibility of technology.

Technological advance brings greater complexity to, among other things, machinery. As machines become larger and more complex, they have the potential to fail because of their very complexity or because their operators do not fully understand their workings and limitations. This mismatch between machinery and operator was not overwhelming in 1900 and beyond but in certain cases produced major catastrophes. As more and more people became reliant on technology, then more and more were likely to suffer if it failed. Put simply, for example, bigger ships meant more people at sea and potentially more casualties in the event of an incident.

Background: The aftermath of the 1906 earthquake and fire that left much of San Francisco in ruins. Some 700 citizens died and a further 200,000 were left homeless.

1900-1909

SAALE, NEW YORK

JUNE 30, 1900

Harbors are meant to be places of safety and refuge for ships, but, as the fate of the *Saale* shows, this is not always the case. This liner caught fire from a conflagration on the pier while the ship was "safely" berthed.

The two-funneled *Saale* was owned by the North German Lloyd line and its home port was Bremen. The ship had accommodation to cater for over 1200 passengers, of whom 150 could enjoy the privilege of first-class travel. The *Saale* worked between Bremen, Southampton, and New York.

In late June 1900, the *Saale* was docked at Hoboken, New Jersey, in the company of three other North German Lloyd ships – the *Bremen*, the *Kaiser Wilhelm der Grosse*, and the *Main*. A fire started on the adjacent pier three among a pile of cotton bales and spread to barrels of oil and turpentine stored nearby. Strong winds assisted the spread of the fire and the docked ships were in danger of being consumed. The *Kaiser Wilhelm* made steam and edged away from the growing conflagration but the other ships caught fire.

Those on the upper deck of the *Saale* were able to escape by leaping into the Hudson River. Unfortunately those below deck were trapped. Some tried to escape through the portholes but these proved to be too small for anyone to get through. The burned-out *Saale* settled on the bottom of the dock.

When the wreck was investigated, it was found that the carnage below decks was horrific. The charred remains of 99 passengers and crew were brought out. One positive result of the disaster was that ships in the future were built with portholes big enough to escape through in an emergency.

The *Saale* was repaired and continued to have an active life for 24 more years, though not as a passenger ship. In 1924 the *Saale*, renamed the *Madison*, was scrapped in Italy.

Below: The German liner the Saale, *which was berthed in Hoboken, New Jersey, when a fire on the adjacent pier spread to the ship, with horrific results.*

PARIS, FRANCE

AUGUST 10, 1903

This incident, which left 84 commuters dead, took place on the French capital's metro system and was caused by a mechanical fire. The Compagnie de Chemin de Fer Métropolitain de Paris's stock on Line Two of the system was powered by electricity, with the power control device situated in the cab at the front of each train. A power line linked the pair of motor cars located at each end of the train. On the evening in question, smoke began billowing from the power controller on Train 43 at the height of the rush hour.

Station staff reacted promptly, evacuating the passengers of the smoking train and calling for the next train to run into the station so that it could move the first to a place of relative safety. Passengers from the second train were also ordered to disembark, thereby adding to the crush on the platforms.

Initially, the intention was to run the damaged train into sidings at Belleville, but this proved impossible because of difficulties with points and it was decided to move the endangered train to the more distant Nation terminus station.

As this procedure was taking place, the leading wooden vehicle burst into flames just as the train entered Ménilmontant station. At the same moment, a following train entered the station at Couronnes. Staff here spotted the denser smoke from the burning train in front flooding out of the tunnel and attempted to get the hundreds of passengers out of the station.

Many refused to move, but a sudden failure in the lighting system induced panic on the crowded platforms. Dozens were killed in the ensuing crush; others escaped by walking back through the tunnel to Belleville station.

Left: French rescue workers struggle to reach the passengers trapped underground following the fire caused by a train's electric system. Many passengers died because of smoke inhalation, while others were crushed to death.

IROQUOIS THEATER, CHICAGO, USA

DECEMBER 30, 1903

A total of 578 people were killed by a fire which swept through the Iroquois Theater, Chicago, USA, in December 1903. A huge matinee audience was watching a Christmas show when the auditorium was suddenly engulfed in flames from a fire that had started backstage. One of the most bitter ironies of this disaster was that Chicago had already been razed to the ground once in October 1871, and many of the new public buildings which had risen, phoenix-like, from the ashes had been designed and constructed with fire safety as the highest priority.

The Iroquois had been completed only the previous month. Posters advertising the inaugural Christmas production of *Mr Bluebeard*, a popular entertainment of the period, described the theater as "absolutely fireproof." *Mr Bluebeard* had attracted standing-room only houses throughout the Christmas period, and on the afternoon of the tragedy there was an audience of about 2000 in a theater with seats for only 1600.

When the fire broke out, comedian Eddie Foy, who was on stage at the time, appealed for an orderly exit, but to little avail. The performers and stagehands escaped immediately through the stage door, but there

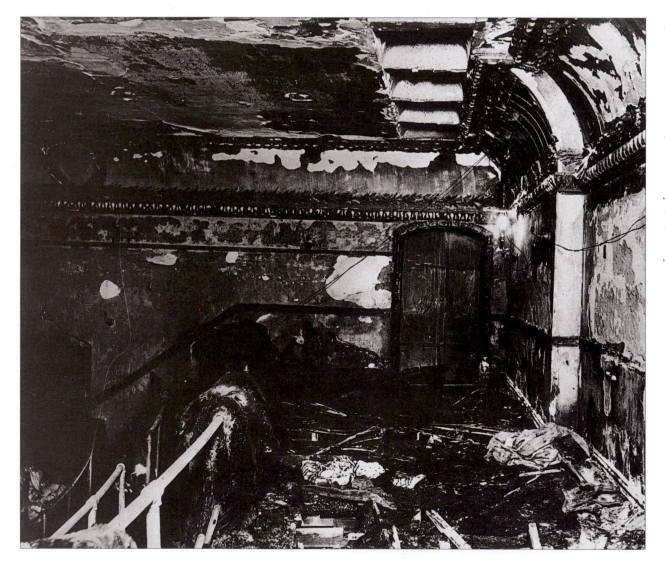

Left: Devastation inside the theater after the fire. The theater had been completed only a month earlier, and its opening production was billed as being a "great spectacular entertainment from the Theater Royal, Drury Lane, London." The theater was filled to capacity at the time of the terrible fire.

Above: The auditorium of the theater was opulently decorated and furnished, and the upholstery on the seating burned fiercely, belching out smoke and fumes that suffocated many of the trapped audience.

were not enough exits from the auditorium to cope with the mass of people. The exits from the stalls led into narrow alleyways, while those from the balcony and circle all joined up into a single passageway, thus funneling people into a crowd that could not disperse in time to avert disaster.

Inevitably, a great crush ensued and the doors soon became jammed with terrified people who panicked in the scramble for safety. Some jumped from the balcony into the pit; others jumped from external fire escapes to their death in the street below. Those who did not reach the exits at all were either burned to death or suffocated by fumes and dense smoke belching out from the upholstery. About 200 people died in this way, but almost twice as many were trampled to death in the stampede to get out – many bodies had boot marks all over their faces.

Fire does its work quickly, and the Iroquois was completely gutted in less than ten minutes. Reputedly the safest public building in Chicago, the theater in fact violated numerous safety laws, including strict rules about overcrowding, and five employees were subsequently charged with manslaughter. Although there was a safety curtain – an essential anti-fire feature in any theater – in the emergency it did not perform its intended function of cutting off the auditorium from the stage area where there are always numerous flammable materials such as greasepaint and wooden scenery. On the afternoon of the fire, the safety curtain seems to have jammed briefly and then been lowered, but only after flames had crossed the threshold between stage and auditorium.

In the aftermath of the blaze, 50 theaters elsewhere in the USA were closed for safety reasons.

GENERAL SLOCUM, NEW YORK

JUNE 15, 1904

Every shipwreck involving the loss of life is a tragedy, but the destruction of the sidewheel ferry *General Slocum* and the huge loss of life that accompanied it had a profound impact on the American people. The disaster seemed the more terrible because the ferry was packed with New Yorkers heading away from the heat of the city to enjoy a summer picnic. A large party of school children from St Mark's School, along with their teachers and parents, were among the passengers.

The *General Slocum* was a large vessel and it was packed to capacity when it sailed out from New York. It was under charter and its destination was Throg's Neck, a popular spot to pass away a pleasant summer's day. As the ferry, commanded by Captain van Schaick, made its way down the East River; those on board were totally unaware that disaster was about to strike. Most of the passengers watched the New York skyline pass, enjoying the cool breeze generated by the ship's progress down the river.

Eyewitness reports differ as to the source of the fire that engulfed the ship. Everybody seemed to agree that it started below deck. Some witnesses stated that the fire began in the ship's galley, the source being a stove; others reported that it started in a paint store. Whatever the truth of the matter, the fire spread rapidly throughout the wooden ship.

Within minutes the ship was in difficulties, but the crew members detailed to fight the fire stuck to their impossible task for nearly one hour. There was no saving the vessel, however, and those caught on board faced a terrible dilemma. They could stay with the *General Slocum* and hope the flames could be brought under control, or they could take their chance in the waters of the East River. Some jumped for their lives, but many of these drowned. Most stayed with the ship and perished in the flames.

The vessel – what was left of it after it the superstructure had been burned to the waterline – eventually sank. The final tally of dead was 1021 passengers and crew – but the true figure may have been higher than this as no one really knew how many people were actually on board.

A disaster of such magnitude was, of course, investigated by the authorities. Captain van Schaick had survived the inferno and played a key role in the investigations. When the board of inquiry delivered its verdict, he was singled out as being at least partially responsible for the catastrophe. He was subsequently tried and convicted on a charge of manslaughter, and received a jail term. So great was the grief of ordinary New Yorkers that the city erected a plaque to commemorate the sinking of the *General Slocum*.

Above: Horrified passengers jump from the burning General Slocum.

SAN FRANCISCO FIRE, USA

APRIL 18–21, 1906

The earthquake that shattered San Francisco, California, USA at 0500 hours on the morning of Wednesday, April 18, 1906, was followed by a devastating fire that burned for four days and almost totally destroyed the city center.

What became a mighty inferno began as dozens of small fires scattered throughout the city. Some broke out from heaters left burning in buildings that had been abandoned during the earthquake. Others spread from hearths and kitchen ranges that had been overturned by the tremors. Many were triggered by severed electric cables or the ignition of gas from fractured gas pipes. Altogether, there were 52 of these small fires. Even in normal circumstances, the San Francisco Fire Department would have been hard pressed to cope with all of them at once because it had only 38 horse-drawn fire engines at its disposal. But these were not

Above: *The devastation in the center of the city was almost total, leaving over 200,000 people homeless.*

Left: *People who managed to escape from the city watched the fire from the safety of Russian Hill.*

Above: At the junction of Mission Street and Third Street rubble caused by the earthquake covers the ground, while, in the background, the Opera House can be seen in flames.

normal circumstances – the city was already in chaos from the earthquake and great fissures in the streets had severed every water main. The only source of water for hoses was the Bay, and not all the fires were along the waterfront.

Fanned by a stiff breeze, the small fires came together in a great conflagration and by noon on the Wednesday it was out of control. Federal troops were called in, along with the National Guard and 600 helpers from the University of California at Berkeley on the other side of the Bay. They did not even try to put out the existing blaze – that was clearly impossible. Instead, they concentrated on saving as many lives as they could. They also dynamited buildings in an attempt to create an unbridgeable gap in the path of the advancing flames, but this turned out disastrously – the buildings blew outwards instead of collapsing and started new fires.

Looting became a serious problem. Martial law was imposed and the army was called in to help the overstretched police force. Mayor Schmitz sent out leaflets warning that thieves would be shot on sight, and at least three people were killed in this way.

By Saturday, the fires had begun to burn themselves out and rain helped to extinguish the last ones. Only then did the true scale of the disaster become clear – about 700 people had been killed, including 270 mental patients locked in their asylum who had burned to death. One man trapped inside a burning building persuaded a policeman to shoot him. City Hall was turned into a makeshift mortuary. More than 200,000 people were homeless and had to camp in tents in Golden Gate Park and other open spaces. Four square miles (10 sq km) of the city had been obliterated, including 514 blocks containing 28,000 buildings. The whole of Chinatown was reduced to ashes, as was all but one house in the fashionable Nob Hill residential district. Losses from the fire were 20 times greater than the losses from the earthquake and led to the largest ever fire insurance claim of 5,748 million dollars.

SALISBURY, ENGLAND

JULY 1, 1906

This collision on the London and South Western Railway occurred shortly before 0200 hours one mid-summer morning, when the express boat train traveling from the port of Plymouth in southwest England to Waterloo station in London failed to reduce speed while negotiating the sharply curving track at Salisbury station in Wiltshire. The boat train, which had been introduced just two years earlier to speed travelers to the capital, was carrying people who had crossed the Atlantic on a vessel owned by the America Line. At the time, it was the only train on the intercity line that did not stop at Salisbury station.

The boat train's locomotive and its passenger coaches left the track and crashed into a milk train passing through the station in the opposite direction. Railroad regulations of the time were meant to ensure that the Plymouth–Waterloo express passed through Salisbury at no more than 30mph (48km/h). Indeed, other regulations indicated that the train was supposed to negotiate the length of track between the station's two signal boxes at no more than 25mph (40km/h).

On this early July day, however, its speed was much higher, closer to the average for the complete high-speed journey from Plymouth to London. The violent impact resulted in 24 of the boat train's 43 passengers being killed, with many others suffering serious injuries. All the fatalities in the boat train were located in its three central cars and only five of its carriages remained undamaged following the smash.

Subsequent investigations revealed that the boat train's driver had attempted to negotiate the curving section of track at Salisbury station at a speed in excess of 60mph (96km/h). This was more than twice the official speed and was sufficient to topple the locomotive and bring it into contact with the milk train with such disastrous consequences. The board of

Below: Rail staff begin to clear the wreckage of the boat train from Salisbury station.

Right: Two steam-powered cranes start to clear some of the overturned debris from the main line through Salisbury.

investigation discovered that railroad companies at the time did not publish speed restrictions, such as the one supposedly in force at Salisbury, in the timetables available to their staff as a matter of course, and suggested that they should be introduced urgently, along with fitting speedometers in locomotives. The use of speedometers in trains did not become universal for almost another 50 years, however.

Passengers injured in the crash and the relatives of those killed received compensation of more than £23,000 some 12 months later, after the board of inves-

tigation had delivered its findings. However, one of the largest settlements, £7000, was agreed only in 1909 and went to a female traveler on the boat train.

The damage to the two trains themselves, however, was perhaps not as severe as might have been expected. Several of the worst effected passenger carriages on the express were scrapped following the investigation, but the locomotives themselves, some of the least damaged rolling stock, and track and other parts of Salisbury station were all repaired at a cost of a little over £5000.

Right: Railroad staff attach heavy lifting chains to the engine of the boat train prior to it being removed from the scene of the accident.

ELLIOT JUNCTION, SCOTLAND

DECEMBER 29, 1906

This accident took place in the depths of winter and was brought about by heavy snowstorms which affected the operation of a local signal system. Twenty-two people died in the subsequent collision between a local commuter train and an express.

The incident began when a southbound freight train was brought to a halt and split into two parts by a snowdrift. Its driver then decided to switch lines and return to Elliot Junction with the portion of the train still attached to his locomotive. Once there, he intended to switch back on to the southbound line, return to the stationary portion of the train and then recouple the two parts.

The maneuver took some time because of the conditions, and some of the freight wagons were derailed during recoupling. With one line closed because of the derailment, rail staff implemented single-line operating. However, heavy snow on telegraph wires put the signal system out of operation. By 1500 hours, the staff at Arbroath station, north of Elliot Junction, began getting their delayed trains under way.

First to leave was a local commuter train heading south and this was followed some 15 minutes later by an express from Arbroath that had been stuck at the station since the morning. Its driver was warned to keep his speed down on three separate occasions by rail staff, but failed to do so. He collided with the local passenger train at a speed of 30mph (48km/h) while it was halted at Elliot Junction waiting to be switched on to the single-track section so that it could pass the derailed freight train.

Below: Wreckage covers the tracks at Elliot Junction after the late December crash. Although the speed of the collision was not particularly high, the damage was extensive.

DAKOTA, PACIFIC
MARCH 7, 1907

The *Dakota* was the pride of the US Great Northern Steam Ship Company. At over 20,700 gross tons, the *Dakota* was 630 feet (191m) long, and had accommodation for 2700 passengers. The ship was the biggest US-constructed passenger vessel when it was commissioned in 1905, a record that was to stand until the late 1920s. However, the *Dakota*'s career was destined to be brief.

The *Dakota*, along with the *Minnesota*, was the brainchild of the owner of the Great North Railroad, James Hill. He saw a gap in the market on the United States–Far East run. Both the *Dakota* and its sister ship the *Minnesota* were built to transport the passengers and freight being carried on his railroad to and from the Far East. As a concession to the tastes of his Oriental passengers, Hill ordered an opium den to be built on both ships.

The *Dakota*'s life came to an end on March 7, 1907. The ship had sailed from the Pacific Northwest and was heading for Japan. It never completed the journey. Some 40 miles (64km) out from Yokohama, the main port of Tokyo Bay and the first Japanese port opened to Westerners in the 19th century, the *Dakota* struck a submerged reef. The ship was stuck fast and the passengers and crew were able to abandon it without mishap. A storm on March 23 put paid to all hopes of saving the ship, however. It broke up during the pounding and was sold for scrap.

The captain, whether out of guilt or shock, gave up his life at sea, and took a job for the remainder of his life as a guard in a San Francisco shipyard.

Right: *One of the last photographs taken of the* Dakota *as it sinks 40 miles (64km) off Yokohama.*

SHREWSBURY, ENGLAND

OCTOBER 15, 1907

The outcome of the investigation into this incident on the London and North Western Railway caused considerable consternation among the system's drivers. They believed that the investigators' decision to blame the driver for the derailment of a 4-6-0 "Stephenson" engine and its carriages was far from watertight and that there may have been other explanations.

Below: The aftermath of the crash. The investigators put the blame on the driver but their evidence was far from convincing.

On the day in question, the train was approaching Shrewsbury station from the direction of Hadnall and was supposed to reduce speed to negotiate a steep bank down to Shrewsbury. As the train was running at night, its driver was meant to take notice of a speed reduction instruction. Nevertheless, the Shrewsbury-bound train did not make any reduction in its speed

and, as it reached a sharp curve in the track over a junction a little way outside the station, was derailed and thrown on to its side. Carriages and the engine were badly damaged, and 18 of those on board were killed. Among the dead were the crew of the engine and three sorters working for the Royal Mail.

Investigators were obviously unable to interview those whose direct involvement in the crash might have been able to shed some light on the matter, yet announced that the driver probably dozed off, thereby causing the incident.

It was clearly a far from complete analysis, and paved the way for considerable speculation in the press. Some newspapers suggested that the newness of the locomotive was responsible, but produced no evidence to support their claims.

The optimism that greeted the dawn of the twentieth century did not survive long. Technology, which many had seen as a cure-all for every problem confronting humankind, was rapidly shown to be a force as much for evil as for good. World War I revealed that it could be harnessed not only for good but also for violent intent. However, faith in technology was shattered before the war began in August 1914, well before humankind's hopes had been crushed in the trenches of the Western Front and elsewhere.

If one event can be said to have fundamentally changed our recent ancestors' belief in technological progress, it was the sinking of the *Titanic* in 1912, which was accompanied by heavy loss of life, and heralded a new relationship with technology. The luxury liner, considered the pinnacle of human achievement, was brought low by a combination of nature, an iceberg, and human arrogance or fallibility, depending on your point of view. The vessel, for all its outward sophistication, was fatally flawed, despite the firm beliefs and claims of its creators. It was ill-provided with lifeboats and its watertight compartments were poorly designed. Its supposed "unsinkability" proved to be a cruel illusion, one that resulted in more than 1500 deaths among the passengers and crew. The sense of loss was overpowering, yet it was but one of many catastrophes between 1910 and 1919 that marked a radical change in our relationship with technology.

Background: *The news of the loss of the* Titanic *in April 1912 is made public. The liner's sinking provoked outpourings of emotion around the world.*

1910–1919

TRIANGLE WAIST FACTORY, NEW YORK, USA

MARCH 24, 1911

One hundred and forty-six workers were killed when fire swept through a shirt factory on the top three floors of a 10-story building at the corner of Greene Street and Washington Place in New York City. Some 125 of the victims were young women aged between 16 and 23.

The Triangle Waist Company – the name of the firm referred to the idealized shape of the male torso when wearing one of its shirts – was a sweatshop in a squalid loft building. The tragedy highlighted the plight of blue collar workers who were exploited by having to put in long hours for poor pay in dangerous conditions.

The fire broke out on the eighth floor at 1640 hours, five minutes before closing time as the workers were getting ready to go home. During the week there would normally have been many more people in the building, but it was Saturday and only the Triangle staff – 700 of them, including 600 women – were working because the owners had a rush job on. Most of the women were Germans, Hungarians, Italians, and Russians, recent immigrants who had been hired after the firm had fired nearly all their old workers – most of whom were Jewish – when they became unionized and demanded better working conditions. At about the same time, the building – which had had four previous fires recently –

Below: The Triangle Waist factory building had 10 storys and was classified as fireproof by the New York City authorities.

Above: The scene inside the garment factory after fire swept through the top floors where women were working on a rush job on a Saturday afternoon.

Right: One of the elevators after the fire. The other elevator was not working at the time of the fire.

had been reported to the authorities because of the inadequacy of its exits. However, no action had been taken to improve them.

The flames spread rapidly through the flimsy material on the racks and the offcuts which were piled high on the floor beside the machines. Some of the victims burned or were suffocated by fumes; others leaped 100 feet (30m) to their death down the lift shafts or from upstairs windows – many bodies were found in the street clutching their pay envelopes. The fire brigade tried to catch some of the jumpers in firenets – sheets held outstretched like trampolines by at least three people – but their efforts were futile: one girl jumped and landed safely, but three others followed her down and landed on the same net before she could get off. All four were killed by the impact.

Nevertheless, many other people jumped whom the fire brigade believed they might have saved. One girl who saw the glass roof of a sidewalk cover at the first floor level of the nearby New York University building leaped for it: her aim was right, but she crashed through it onto the sidewalk. Another girl waved a handkerchief as she jumped – about half way down she

Above: Firemen searching for bodies after the fire. Many women jumped to their deaths from the top floors of the building where they were trapped.

got her dress caught on a wire sticking out of the side of the building. The crowd watched helplessly as she hung there until her dress burned free and she fell to her death on the road.

The emergency services were so busy dealing with the fire that they could do no more than pile up dead bodies in the street. Then someone noticed that among them was a girl who was still breathing – they pulled her out at once, but she died two minutes later.

The structure of the building had been classed as fireproof by the New York City authorities. Although its exterior emerged virtually undamaged, inside the building the top three floors were completely gutted in less than half an hour. There was only one fire escape, and that was internal.

Two of Triangle Waist's owners, Isaac Harris and Max Blanck, were in the building at the time of the fire but escaped over the roof. This was an exit unknown to the staff who had always used the two freight elevators, one of which was not working on the afternoon of the fire. Police investigators subsequently discovered that two of the girls who had died, who were working on the ninth floor, had been locked into their work room. US labor law had strict rules forbidding employers to lock or bolt in their workers.

The Triangle Waist fire and its causes were important in the growth of the US labor union movement. Previously, clothing unions had struggled to attract new members because workers were too worried about keeping their jobs to agitate for better pay and conditions. This fire drew public attention to the plight of people in sweatshops; and in reaction against the fire, large numbers of garment workers now joined labor collectives.

TITANIC, ATLANTIC

APRIL 15, 1912

As well as being a major maritime disaster, the loss of the *Titanic* was also a highly symbolic event. The 46,000-gross ton liner was built at a time when countries gained great international prestige from these marvels of engineering. Apart from the considerable amount of money to be made from the lucrative transatlantic trade, the great liners came to symbolize a country's wealth and international status. The *Titanic* was the latest in a series of luxury liners that were intended to convince their passengers that they were as safe on sea as they would be on land. The loss of the *Titanic* was a signal that the sureties of the age were built on foundations of sand.

Below: When it was launched in 1911, the Titanic *was the latest thing in luxury liners and was widely believed to be unsinkable.*

The *Titanic* was born when the managing director of the White Star Line, J. Bruce Ismay, decided to build three vessels that in size and luxury would outstrip anything else afloat. The first, the *Olympic*, was launched in 1909, and gave distinguished service. The third, the *Britannic*, saw action in World War I until sunk by a mine in the Mediterranean. However, it was the second ship, the *Titanic*, that was to become one of the most famous ships of all time.

All three ships were designed to carry some 2500 passengers in three classes and had crews of almost 1000. All had what were thought to be state-of-the-art safety features, including 16 watertight compartments. However, as events were to show, faith in these watertight compartments was misplaced.

The *Titanic* was launched in 1911 amid great publicity, and was widely believed to be "unsinkable." White Star planned its maiden voyage from Southampton to New York, via Cherbourg and Cork, for April 10 the following year. When the liner left Southampton, it was commanded by Captain Edward Smith.

Smith was a good choice for the prestigious first voyage. He was very experienced and been master of more than 15 of the line's ships. Despite his experience, however, the voyage began badly. As the *Titanic* left Southampton, it displaced a huge volume of water which dragged a second liner, the *New York*, from its moorings. A collision was only narrowly avoided. Although they did not know it, the 1308 passengers (who included Ismay, the managing director of the shipping line) and 898 crew on board had had a warning of the fate of the *Titanic*.

The *Titanic* stopped off in Cherbourg and Cork and then headed out across the Atlantic. There were reports of icebergs, but they were to the north of the route being followed by the *Titanic*. Ismay was eager to prove his new liner's worth, and although he had no right to do so, Ismay ordered Smith to steam ahead at full speed. The *Titanic*'s engines performed well. In the first day the ship made an impressive 546 miles (873km). Radio reports of icebergs farther to the south continued to be logged, but there was no reduction in the liner's 22 knots.

As night fell on April 14, lookouts were posted around the vessel to keep careful watch for icebergs or other dangers, and the captain went to his cabin. Shortly before midnight one lookout spotted an object ahead of the liner. It was an iceberg and he immediately reported the danger. His action was prompt but the liner was traveling at high speed. The ship's rudders were put to port and the engines full astern, but the *Titanic* was a large vessel and did not respond quickly enough. The ship struck the iceberg.

Those on board later reported nothing more than feeling a jolt and hearing some scraping noises, but the collision had fatally wounded the vessel. There was no panic – after all, the ship had been labelled "unsinkable." However, there was a fatal design flaw. The watertight bulkheads did not reach the full height of the vessel. As one watertight compartment filled, water flowed over the top and filled the next compartment, and so on. Thomas Andrews, a passenger who was also managing director of the ship's builders, told the captain the *Titanic* would sink in about two hours.

Smith ordered the sending of the new SOS distress signal and one ship, the *Carpathia*, picked it up and steamed toward the *Titanic*. It was just 50 miles (80 km) away from the liner. Help was closer at hand, however. Another ship, the *Californian*, was reportedly just a few miles away but its radio operator was off duty and the *Titanic*'s distress signals went unheard. On board the *Titanic*, Smith gave the order to abandon ship. However, there were just 20 lifeboats for 2200 passengers and crew. The builders had believed

Left: It was just before midnight on April 14, 1912, that the Titanic *struck an iceberg in the North Atlantic.*

Right: After the ship struck the iceberg, some lifeboats were lowered and first-class passengers were taken off. But there were not enough boats to take all the passengers and the ship's crew.

Below: As the Titanic *starts to sink, survivors in a lifeboat battle it out against heavy seas and freezing weather conditions.*

that, even in the worst scenario, the *Titanic* would stay afloat long enough for rescue ships to arrive. Lifeboats would only be needed, it was thought, to shuttle passengers from the *Titanic* to the waiting vessels. This proved to be a fatal conclusion.

There was little immediate panic among the *Titanic*'s passengers. Women and children were hurried into the lifeboats, but the class system prevailed. Those who were saved were almost all first-class passengers. The vast majority of those who died – men, women, and children – were in the third class. Ismay survived but Smith went down with his ship. The *Titanic* slid beneath the waves at about 0200 hours on April 15. The *Carpathia* arrived on the scene about two hours later, and picked up 703 people. The final total of the dead was 815 passengers and 688 crew.

The sinking of the *Titanic* made headlines around the world and questions were asked about how such a fate could befall such a great liner. The White Star Line was partly a victim of its own publicity – it had been so sure that the liner's safety features made it invincible. Boards of inquiry on both sides of the Atlantic, however, pointed out a number of shortcomings. There were not enough lifeboats to accommodate the

number of passengers and crew carried – even though according to the guidelines in place at the time, the *Titanic* did have the correct number of boats. No boat drill had been carried out. And the ship was traveling at high speed at night in a known iceberg area.

The loss of the *Titanic* led to new safety regulations at sea. Ships were to carry enough lifeboats to accommodate all passengers and crew, and boat drills were to be held on all voyages. Vessels were required to follow a more southerly route across the Atlantic, and an ice patrol system was introduced to give advance warning of icebergs. And all ships were in future to keep a round-the-clock radio watch in case of distress signals.

Left: The scene on board the sinking Titanic, *as women and children are herded on to the lifeboats, leaving their menfolk behind.*

Below: The last moments of the Titanic *as the great ship slides beneath the water. Legend has it that the ship's orchestra continued to play as the ship went down.*

COLCHESTER, ENGLAND

JULY 12, 1913

This fatal collision in the East Anglian town could have been much worse except for the quick reactions of the driver of the Harwich train, one of two involved in this rear-end collision. Nevertheless, a large amount of rolling stock was smashed and several rail employees were killed when an express heading for Cromer on the north Norfolk coast hit the Harwich train at speed.

Below: The remains of the trains involved in the fatal collision at Colchester.

The Harwich goods train reached Colchester at 1411 hours, uncoupled, took on water, and was put at the disposal of the station for further duties. Because of the presence of a second train, the Harwich locomotive had to wait on the main line before being switched.

Unfortunately, due to a signalman's error, the Cromer express was allowed to continue on the same line as the stationary Harwich train, rather than waiting until the line ahead was cleared.

The Harwich train's driver sensed that something was wrong and sent his fireman to the signal box. At that moment, he spotted the Cromer express approaching at 40mph (64km/h) and attempted to get his locomotive under way. However, his train was struck and pushed nearly 300 yards (277m) along the track and then fell on its side. The express locomotive was badly damaged, with only its boiler being salvaged, and 14 people were injured, including the quick-thinking driver of the Harwich train.

SENGENHYDD MINE, WALES

OCTOBER 14, 1913

Britain's worst coal mining disaster was caused by the fire which followed a gas explosion deep below ground at the Universal Colliery, Sengenhydd, South Wales. Although 500 miners were brought up alive, 439 others were killed. The initial blast was so loud that – despite being underground – it was still clearly audible 11 miles (18km) away in Cardiff.

From the mid-19th century onwards, when coal became the lynchpin of the British economy, Wales had a reputation for having the best seams and the worst mines. The pits were dangerous partly because they were dug so deep and partly because the seams they exploited had a high content of explosive gas, particularly firedamp, a form of methane.

Although the Sengenhydd mine was a comparatively recent working – it had been fully opened only in 1896 – it had already been the scene of a major tragedy: in May 1901 a large methane explosion had killed all but one of the 82 men working below ground. Nevertheless, the mine was also one of the most

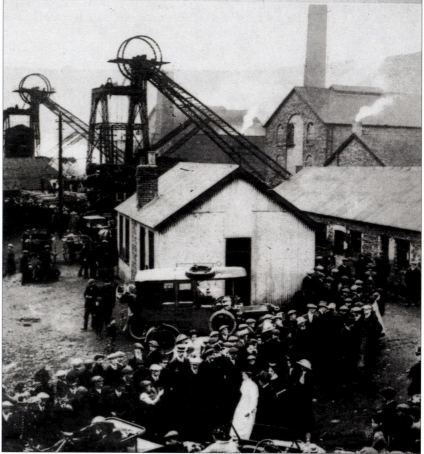

productive for its size: by 1913, Sengenhydd was producing about 1800 tons of coal every day.

Because of the earlier tragedy, the Sengenhydd pit was examined regularly for firedamp, and indeed it was checked first thing on the morning of Tuesday, October 14, before the early shift went down at 0600 hours. At 0810 hours, however, a huge explosion ripped through the mine and blocked the adit (entrance) with rock and debris. The pit's own rescue crew cleared the rubble and went down immediately, but when they reached the main underground working, they found that half the pit – and half the miners – were cut off by a wall of fire.

Although the rescue teams quickly realized that they needed outside assistance, the fire brigade was unaccountably not called out for nearly two hours and the long delay is believed to have been one of the major contributory factors to the extent of this tragedy.

By the end of the day, rescue teams had found no survivors and only 12 bodies had been removed from

Above: Anxious families gather at the pit head as news of the disaster spreads around the mining village.

Far left: The scene at Sengenhydd mine after the explosion that fatally trapped 439 miners.

Right: The underground explosion ripped apart the gear at the head of the pit. Immediately afterwards one half of the mine erupted in a mass of flames.

the mine. The following morning the search resumed and 18 men were found alive in one small area. Later, they pulled out an unconscious boy who was revived after two hours of artificial respiration.

It was several more days before all the miners who had been down the pit were accounted for. When the extent of the tragedy became known, grief swept the nation. There was a government inquiry which found that, despite numerous safety precautions, the mine remained appallingly dangerous. Seam walls and ceilings which should have been swept regularly were covered in flammable coal dust. Newly introduced signalling equipment was run by batteries and the sparks from them could have ignited the firedamp. The steam fan which circulated air through the mine was also a hazard and possibly one of the causes of the tragedy because, unlike the ventilation systems in other pits, it could not be instantly reversed to stem the flow of flame-fanning air. The Sengenhydd Colliery manager was eventually brought to court and fined £24 – local people bitterly remarked that this was about one shilling per dead miner.

EMPRESS OF IRELAND, ATLANTIC

MAY 29, 1914

The *Empress of Ireland* was a comfortable, if unostentatious, passenger liner that was built by Glasgow's Fairfield Shipbuilding and Engineering Company in 1905 for the Canadian Pacific Line. The *Empress of Ireland* plied the Quebec to Liverpool transatlantic route, and the liner's final voyage began on May 28, 1914. The vessel sailed from Quebec with 1057 passengers on board and Captain Henry Kendall in charge of 420 crew members. The ship also carried about 1100 tons of general cargo.

As the liner reached Father Point, the pilot disembarked and the *Empress* continued on its journey to Liverpool. As the ship pushed on lights were spotted at about a distance of six miles (9.5km). They were from the *Storstad*, a Norwegian collier with 11,000 tons of coal in its holds. As the two vessels closed both seemed to believe they would pass each other with room to spare. However, as they closed, thick fog covered the area, reducing visibility. Those on watch tried to sight the other approaching ship by its lights.

The *Storstad* spotted the *Empress* first and made a desperate attempt to turn away. Kendall first ordered full ahead and then astern to avoid the *Storstad*, but it was far too late. The *Storstad* struck the *Empress*, opening up a large hole in the *Empress*'s hull through which seawater poured in.

Kendall realized that the *Empress* was finished and, after an unsuccessful attempt to get the ship beached, he gave the order to abandon ship. However, most of the passengers were asleep, and were slow to respond. The *Empress* keeled over on to its side and went under bow first less than 15 minutes after the collision. Of the 1057 passengers aboard 840 drowned, only 217 surviving. The crew were luckier – or more prompt to react – and 248 of the 420 were able to escape.

Although the loss of the ship did not receive anything like the same news coverage as the loss of the *Titanic*, in fact more passengers perished from the *Empress* than from the *Titanic*.

Right: Recovering bodies from the St Lawrence River after the wreck of the Empress of Ireland.

LUSITANIA, ATLANTIC
MAY 7, 1915

The sinking of the *Lusitania* in the second year of World War I was one of several incidents involving the loss of American lives that eventually led to the United States declaring war on Germany in 1917. The *Lusitania* was the first of the Cunard Line's great luxury liners and when it won its owners the Blue Riband for the fastest transatlantic crossing in 1909, the *Lusitania*'s future seemed secure. The crossing was completed at an average speed of just under 24 knots.

Although primarily a luxury passenger liner, the *Lusitania* had the potential to be much more. The ship had been designed in close cooperation with the British Admiralty. The Admiralty, suspecting that war

with Germany was almost inevitable, demanded that Cunard build a fast ship laid out in such a way that it could be fitted with armaments. War was still some way off when, in May 1913, the *Lusitania* was given a secret refit – its port and starboard shelter decks were modified to take two four-gun batteries and room was made for two magazines to store the guns' ammunition. War was declared in August 1914 and by mid-September the *Lusitania* had been designated by the British Admiralty as an armed auxiliary cruiser.

The *Lusitania* continued to sail between Liverpool and New York as a passenger liner, despite Germany's declaration that the waters around Great Britain were a war zone and that any ship flying the British flag was likely to be sunk.

Above: The first of the Cunard Line's great luxury liners, Lusitania set out on her maiden voyage from Liverpool, England, on September 7, 1907.

Left: The first-class public rooms of the Lusitania – such as the lounge shown here – were decorated in an opulent Edwardian style to encourage the passengers to think they were in the safety of a luxury hotel on dry land.

The *Lusitania*'s final voyage began on May 1, 1915. The vessel was sailing from New York to Liverpool. On board were over 1150 passengers and 700 crew members, and 1400 tons of "general" cargo. Packed into the ship's holds, this included over 1200 cases of artillery shells and nearly 5000 boxes of cartridges.

Most of this cargo was positioned next to the bulkhead leading into the No. 1 boiler room, which had been converted into a magazine in 1913.

The ship's master, Captain William Turner, encountered no problems on the first part of the voyage back to Liverpool, and the ship made good time. However,

Left: The first-class dining room on the Lusitania.

Above: The Lusitania *went down less than 20 minutes after the first torpedo struck. Almost 1200 people perished.*

on May 6, the Admiralty in London began to send out warning messages regarding German submarine activity in the Irish Sea, which the *Lusitania* had to sail through to reach its home port.

Turner set in motion the established safety procedures – the liner's lifeboats were readied for evacuating passengers, extra lookouts were posted around the ship, portholes were covered up, and many watertight doors were closed.

On May 7, Captain Walter Schwieger of the U-boat U-20 spotted the *Lusitania* as it sailed off the Old Head of Kinsale. The liner was, unusually, sailing in a straight line rather than zigzagging. Shortly after 1400 hours the U-20 launched a single torpedo, which hit the *Lusitania* on its starboard side close to the bridge and near the bulkhead leading into the No.1 boiler room. Water flooded into the starboard coal bunkers and the ship took on a list. The *Lusitania* was then rocked by a second explosion caused by the detonation of its cargo of ammunition.

There was little panic among the passengers, although the ship's list to port made it impossible to launch the boats on that side. When the passengers rushed to the ship's starboard side, they found there were not enough lifeboats to accommodate them all. The *Lusitania* sank in less than 20 minutes from the first explosion, and many people went down with the ship or were lost in the sea. Over 700 passengers perished and only 289 of the crew survived the ordeal. Among the losses were 124 US citizens. The news of

Far right: Half the Lusitania's *lifeboats could not be launched, and there was not enough room in the remaining boats for everybody on board.*

Above: Two decades after the sinking, the wreck of the Lusitania *was found and photographed on the seabed by an American diver.*

the sinking of the *Lusitania* with the loss of civilian lives was blazed across newspapers around the world.

Both the US and British governments investigated the sinking of the liner. The US board of inquiry concluded that the loss was "an illegal act of the Imperial German government." However, the German authorities responded that the *Lusitania* had been warned of the dangers of sailing in Irish waters, that it was an auxiliary cruiser, and carrying war goods and, most remarkably, Canadian troops.

The British conceded that the liner had been converted to carry armaments, but said it was not doing so at the time of the sinking. Surviving passengers claimed to have seen no evidence that the ship was mounting guns or carrying Canadian troops. Later commentators speculated that the British government was actively seeking the destruction of the *Lusitania* as the loss of US lives would bring the United States closer to war with Germany. But no concrete evidence has ever been presented to support this theory.

QUINTINSHILL, SCOTLAND

MAY 22, 1915

Above: A news photograph taken a short time after the incident at Quintinshill gives a good indication of the effort required to re-open the lines blocked by the collision.

This accident in World War I remains the worst railroad accident to have occurred in Great Britain. The collision of several trains resulted in 277 deaths and caused a further 246 injuries; the two employees of Caledonian Railway found responsible for the crash were convicted of manslaughter and given prison sentences.

The tragedy at Quintinshill began with the late-running of two overnight expresses that were heading from London's Euston station to Glasgow. A little after 0600 hours, a third train involved, a stopper from Carlisle, was given permission to head along the main intercity track in advance of the first express.

It was planned to take this Carlisle train off the main line at Quintinshill, which lay 10 miles (16km) to the north. Quintinshill had two side loops, one for use by up trains to London and one for down traffic to Scotland, on to which trains could be diverted from the main line. On the eve of the disaster, the down loop at the station was already holding a freight train heading north, so it was decided to reverse the Carlisle train over to the up line.

It was at this point that the two signalmen enter the story. They had adopted an unofficial shift system that required the nightshift worker to use a separate piece of paper to record train movements after 0600 hours. These would then be entered into the official register by the dayshift signalmen, once he had arrived from Gretna Green Junction on a local stopping train. Shortly after the arrival of the Gretna train but before the replacement signalman had taken over from the nightshift worker, an empty coal train heading south was permitted to make for Quintinshill and wait on the up loop there because Carlisle station to the south could not cope with it due to the weight of traffic using the line there.

The nightshift signalman also accepted the first of the London–Scotland expresses into his zone of responsibility at this point and, while he was having a discussion with his replacement, was also informed that a special troop train was on its way south from

Larbert to Liverpool and could be expected to arrive at any moment.

The dayshift worker then took over the signal box which also contained two men from the trains occupying the loops and the fireman from the Gretna stopper, who, as regulations required, reported the presence of his train. The fireman returned to his train after a brief discussion in the signal box. One of the signalmen there then sent a signal which gave the all clear to trains on the up route.

The troop train was permitted to continue on the line and smashed into the local train, to be followed by the first of the London–Scotland expresses which smashed into the wreckage across the track. The violence of the collisions caused raging fires to break out that made many of the bodies unrecognizable.

Left: Rescuers deploy a lone fire hose to damp down the still-smouldering carriages on the morning after the collision. The ferocity of the fire can be gauged by the coach in the foreground.

Left: The dead and injured from the incident are gathered together in a field adjacent to the crash site. Many of the bodies were burned beyond recognition.

ANCONA, ATLANTIC

NOVEMBER 8, 1915

Passenger liners sailing in time of war are at risk, particularly if they are owned by a line from one of the warring nations. The tragedy of the Italian-owned *Ancona*, which was sunk by a German submarine, was that Italy was not at war with Germany in November 1915.

The *Ancona*, owned by the Genoa-based Italia Società di Navigazione a Vapore, set sail on its last voyage from Naples on November 6 heading for New York. On the 8th the passenger ship was spotted by a submarine which, although actually German, was flying the Austrian flag. The submarine gave chase and slowly closed on the unsuspecting *Ancona*. The captain of the submarine could have used his deck gun to bring the

ship to a halt, and allow the passengers and crew to escape before sinking the vessel. Instead, the submarine fired torpedoes at the defenseless liner.

The *Ancona* was hit at about 1300 hours. The damage was extensive and there was no hope of saving the vessel. The order was given to abandon the ship, which sank rapidly. Many made good their escape but 194 people died. The survivors were finally picked up by a French warship, the *Pluton*.

Eleven US nationals were among the dead and their government demanded an explanation from Austria. It then became clear that the submarine had in fact been a German boat, the *U-38*. This episode hardened US opinion against Germany and brought the United States closer to declaring war.

Above: The Italian passenger liner the Ancona, *which was torpedoed and sunk by a German U-boat in November 1915.*

MONT BLANC, HALIFAX

DECEMBER 7, 1917

By any standards, this incident ranks of one of the worst shipping disasters of the 20th century. Although it occurred during World War I, the destruction of the *Mont Blanc* – together with other ships and a sizeable part of the Canadian port of Halifax – owed nothing to enemy action.

The French cargo ship, the *Mont Blanc*, left New York early in December 1917 carrying a cargo consisting of TNT, benzene, picric acid, and gun cotton – an unstable and potentially lethal mix. The *Mont Blanc* sighted Halifax harbor at 0900 hours on December 6.

Halifax is a natural deepwater harbor stretching for several miles. It is mostly more than one mile wide but narrows to just under half a mile at one point. As the *Mont Blanc* was being guided through the narrows by a pilot, crew members spotted a Belgian freighter, the *Imo*, sailing towards them. The *Imo* should have passed the *Mont Blanc* to starboard but signaled that it would pass to port. The *Mont Blanc*'s captain hauled his rudder over to avoid the other ship, but it was too late. The *Imo* crashed into the *Mont Blanc* close to the hold were the picric acid was stored.

The *Imo* was able to reverse away from the collision but was left unmaneuverable. Fire broke out on the *Mont Blanc*. Its crew abandoned the ship, and it began to drift toward Halifax. Some of the inhabitants of Halifax knew what cargo the *Mont Blanc* carried and how dangerous it was – and ran for their lives. A party from a British cruiser, the *High Flyer*, made an attempt to board the *Mont Blanc* and scuttle it. As they approached by ship's cutter, the *Mont Blanc* exploded.

The initial explosion was massive. Reports suggested that half of Halifax was flattened by it and the fire that followed. The impact of the explosive force was heightened by local geography. The land rose rapidly from the harbor, containing the force of the blast in a small area. Many were killed at the town's railroad station, which was wrecked. The homes of the dock workers around the harbor collapsed like a pack of cards, burying everyone inside.

The horror of the explosion was matched by the fire that followed. It spread rapidly across Halifax, engulfing two nearby areas, Dartmouth and Richmond. Another vessel, the *Pictou*, was moored in the harbor. It, too, carried ammunition. Remarkably, the *Pictou* survived the destruction of the *Mont Blanc* and the likelihood of a second explosion from its cargo was averted when the freighter's sea-cocks were opened and the ship flooded.

Below: The Imo *aground in the landlocked harbor of Halifax after the explosion of the* Mont Blanc.

Right: The railroad station at Halifax was wrecked by the explosion.

The final list of casualties was enormous. Whole areas of housing were destroyed and by the end of the day there were an estimated 25,000 people left without homes in the depths of a Canadian winter. The number of dead was put at between 2000 and 3000, and some 8000 were treated for a variety of injuries. The center of Halifax was a shambles – most of its wooden buildings had been flattened or consumed by fire.

Right: Most of the timber-built houses of Halifax were destroyed by the blast from the explosion. Many of them collapsed, crushing all those trapped inside.

MODANE, FRANCE

DECEMBER 12, 1917

Above: Soldiers and engineers study the remains of the train involved in the Modane incident. So great was the loss of life that the French authorities tried to censor reports of the smash.

This accident, one of the worst in rail history, was a direct result of pressures put on the French system by World War I. The incident began with two trains of Italian rolling stock taking more than 1000 French troops home on leave for Christmas. After passing through the Mount Cenis tunnel, the two trains were coupled together at Modane so that they could negotiate a down gradient hauled by a single train. The train's driver, aware that the total weight of his 19 carriages was more than three times that permitted on the gradient, initially refused to undertake the operation.

Threatened with court martial, he finally agreed to proceed and set off at 6mph (9km/h). Despite applying hand brakes, the train quickly built up speed on the down gradient, partly because of the lack of grip caused by the wet rails. At a point where the line crossed the River Arc, the locomotive separated from the train after one of its bogies had become derailed. The leading coach was also derailed and came to rest against a retaining wall, where it was crushed by the remainder of the train.

Fire broke out and spread quickly, aided by the wooden rolling stock and the explosions of live ammunition that the soldiers were carrying home as souvenirs. Rescuers were kept at a distance from the wreckage because of the explosions. Over 540 soldiers were killed on the troop train, whose driver was later found not guilty of negligence by a court martial. News of the disaster was kept from the French public by the wartime censors.

WEESP, NETHERLANDS

SEPTEMBER 13, 1918

Above: The remains of the train derailed at Weesp lying at the foot of the embankment that gave way. The ship is loading casualties to take them to hospital.

This fatal derailment, which left 41 passengers dead and over 40 others seriously injured, was brought about by heavy rains that weakened the structure of an embankment. On the day in question, an express train of 11 carriages was traveling along a high embankment prior to crossing the Merwede Canal. Unbeknown to the rail authorities a prolonged period of heavy rain had loosened the sand from which the embankment had been constructed.

As the train passed along the embankment, its vibrations were sufficient to set the saturated sand in motion. The train was derailed and its locomotive crashed into the steel girders of the canal bridge. Its wooden carriages were badly smashed as they tumbled down the embankment. However, the casualty list could have been much longer except for the prompt action of an army unit which rushed to the scene less than ten minutes after the derailment.

Other help for those injured was even closer at hand. A doctor, one of the train's passengers, was also able to administer immediate aid to the injured and vessels working on the canal transported many of the injured to hospital.

Investigators studying the crash site discovered that the sand embankment was far too steep and that a layer of clay had prevented the water from draining away, thereby permitting the sand to become saturated. The rail company was held responsible for the flawed construction of the embankment and ordered to compensate the passengers of the express.

The end of World War I seemed to offer a chance of a better world. Once again it was hoped by some that the bloodletting brought about by technology was over and that science would once again be harnessed to create a better world. Certainly, ships and trains were becoming faster and safer, and the possibilities of flight for both business and pleasure (at least for the rich) were beginning to be recognized. However, the lessons of the recent past were not easily ignored, and people were less willing to entirely forget the potential pitfalls of technological fallibility or the dangers presented by extreme natural events over which humankind had virtually no control.

Events quickly proved that people were right to fear nature. This was confirmed in 1923, when earthquake tremors sparked numerous fires in Tokyo and Yokohama. The event was eerily reminiscent of 1906 San Francisco, and the results were no less catastrophic. At least 150,000 Japanese citizens died and many more were injured. Both cities had few buildings capable of withstanding the shocks and most were built of flammable materials. The firefighting and rescue services available were wholly inadequate. The earthquake highlighted a point relevant to any disaster. An individual event cannot be precisely predicted, either in its timing or magnitude, but those responsible for dealing with its consequences need to have in place the resources necessary to minimize its human impact.

Background: Rescue workers search through the remains of one of the Japanese cities devastated by an earthquake and fires in September 1923.

1920-1929

CITY OF HONOLULU, PACIFIC

OCTOBER 12, 1922

When fire breaks out on a ship at sea it can be a terrifying experience. Fire can quickly take over and engulf even the largest of ships, as the fate of the *City of Honolulu* illustrates.

The *City of Honolulu* began life as the *Friedrich der Grosse*, part of the fleet owned by North German Lloyd. The *Friedrich* was used until the outbreak of World War I on the route between Australia and the North Atlantic. However, North German Lloyd lost the ship at the beginning of World War I when, in August 1914, the first month of the war, it was interned in New York harbor.

By 1917, the year the United States entered the war, the US Navy needed to build up its transport fleet to ferry scores of thousands of troops to the Western Front. The *Friedrich der Grosse* fitted the bill, was renamed the *Huron*, and converted to use as a troop-ship. As the *Huron*, the vessel survived the war and in 1922 was chartered out to the Los Angeles Steamship Company. Renamed (yet again) the *City of Honolulu*, the liner began sailing between California and the Hawaiian Islands in September 1922.

The *City of Honolulu* never completed its maiden voyage. Returning to California, fire broke out some 650 miles (1040km) out from San Pedro on October 12. The fire quickly took over the ship, which was completely gutted. A distress call quickly brought the freighter *West Faralon* to the scene, and all of the 70 passengers and 145 crew on board the *City of Honolulu* were rescued.

The burned-out and abandoned *City of Honolulu* drifted for five days before being sunk by gunfire from, ironically, a US Navy warship, the *Thomas*.

Right: To prevent a panic when fire broke out, the captain of the City of Honolulu *ordered the band to keep playing so that passengers could dance until the lifeboats were ready to launch.*

TOKYO AND YOKOHAMA, JAPAN

SEPTEMBER 2, 1923

One of the most destructive fires of all time broke out after three earthquakes in quick succession devastated the Japanese capital, Tokyo, and the neighboring port of Yokohama, which lies eight miles (13km) to the southwest on Tokyo Bay.

The tremors, which were first felt in the morning at 1150 hours, quickly created huge chasms across the streets of both cities and many people were instantly swallowed up inside them. Overhead electricity cables and telephone wires snapped and fell onto passers-by who were immediately electrocuted. Underground gas mains were severed and then exploded, setting off fires which created even greater devastation.

Driven by strong winds, the flames soon engulfed the greater part of both city centers. In Tokyo, many people tried to find refuge in the open air – two of the most popular refuges were the grounds of the Imperial Palace and the shallow canals, where men, women, and children stood or were held for many hours in the hope that the fire would eventually exhaust itself.

But the flames were so strong that they were not impeded by water and many people were burned to death as they stood in the canals by whirling airborne sheets of flame. In Yokohama, the fire was further fuelled by explosions in the dockside oil tanks that had been ripped open by the earthquake.

The fire raged for 36 hours, and – although the army and the rescue services did their best – nothing could be done to impede its progress. In an effort to create a firebreak, unaffected buildings in its path were hastily

Below: Fire enveloped the center of Tokyo following the earthquake. Many people died in burning buildings and even in the open they were not safe from the flames.

demolished. Even this proved futile – either the explosions started new fires or the existing fire just jumped over the space before carrying on its way.

In the aftermath, although it was impossible to establish the cause of every death, it was generally agreed that the fire had killed more people than the earthquakes. Even the combined death toll remains unknown, but it is believed that no fewer than 150,000 people perished. A further 100,000 people were seriously injured.

The damage to property was so great that no one ever bothered to try and calculate it. At least 700,000 dwellings were destroyed – from the smallest houses to the largest hotels. In Tokyo, 17 libraries were burned to the ground, including the Emperor's priceless collection, as well as 151 Shinto shrines, 633 Buddhist temples, and numerous ornamental gardens.

In Yokohama – where the fire and the earthquake had been almost simultaneous – all the port buildings were destroyed, along with the American hospital and two large hotels. As the second and third shocks quickly followed, many people took to rowing boats and headed out into the Bay, where they thought they would be safer. But the flames pursued them even over the water. Although 12,000 people were picked up by the *Empress of Australia*, a liner docked at a safe distance from the shore, the death toll in Yokohama was at least 21,000.

Below: The shopping center of Tokyo suffered heavily in the earthquake and subsequent fire. Little remains of the Ginza, the "Street of Silver," so-called because of the amount of money that changed hands in the shops.

PRINCIPESSA MAFALDA, ATLANTIC

OCTOBER 25, 1927

Right: As the holed Principessa Mafalda *settles in the water, the lifeboats are launched and the ship's passengers and crew try to escape. Even so, over 300 people perished when the ship went down.*

The luxury liner *Principessa Mafalda* was built in 1909 and could carry up to 1700 passengers (in two classes) and around 300 crew at a top speed of 16 knots. The ship plied the lucrative route to South America, chiefly Buenos Aires, from both Naples and Genoa for the Lloyd Italiano Line until the company was bought by the Italian Navigazione Generale Italiana in June 1918.

The *Principessa Mafalda*'s last voyage was supposed to take it to Rio de Janeiro but the vessel never reached its intended destination. Leaving the Cape Verde Islands on October 8 for the final leg of its journey, the *Principessa Mafalda* had 288 crew and 971 passengers on board. Disaster struck as the vessel was near to Abrolhos Island off the coast of Brazil on the morning of October 25. The shaft of the ship's port propeller broke off with considerable force. The damage caused to the immediate area was considerable, including a sizeable hole torn out of the hull. Water flooded into the engine room and quickly filled the boilers. The build-up of steam caused an explosion that blew the boilers to pieces.

The captain immediately signaled for help, but the *Principessa Mafalda* was already settling in the water and listing heavily to port. Seven vessels heeded the distress call and made all speed for the stricken liner. The *Principessa* survived for just under four hours after the explosion before capsizing, taking over 300 passengers and crew to the bottom.

VESTRIS, ATLANTIC

NOVEMBER 12, 1928

The ill-starred *Vestris* was commissioned by Lamport and Holt of Liverpool, England, and first put into service in 1912. At the time of its launch its owners had no idea that the vessel's loss 16 years later would be partly responsible for the company withdrawing from the lucrative service to New York. The 10,494-gross ton *Vestris* was built to carry up to 610 passengers, including 280 in first class, and had a top speed of 15 knots.

When it was first put into service the *Vestris* worked the route between New York and La Plata in South America, but it was then chartered by the Cunard Line and later Royal Mail Lines before being returned to Lamport and Holt in 1922. The fateful journey began on November 10, 1928. Captain W. Carey took his ship out of New York harbor and headed for Buenos Aires. On board were 197 crew and 129 passengers.

The *Vestris* soon encountered heavy seas due to the worsening weather, and the vessel developed a list. This would not have been too dangerous except that some of the ship's cargo and bunker coal began to shift because of the list. As the list increased, some 300 miles (480km) out from Hampton Roads, Carey ordered the sending of an SOS signal and placed the passengers and most of the crew in lifeboats.

Before all of them could be evacuated the *Vestris* went over, going to the bottom with 68 passengers and more than 40 of its crew. The prompt arrival of rescue ships, including the battleship USS *Wyoming* and the North German Lloyd Line liner *Berlin*, undoubtedly prevented the loss of life from being even greater.

Right: A picture taken aboard the Vestris *a few minutes before the liner sank, with the loss of over 100 lives.*

FORT VICTORIA, ATLANTIC

DECEMBER 18, 1929

The sinking of the *Fort Victoria*, a 7784-gross ton passenger ship, was fortunately not accompanied by any loss of life. However, the ship's loss in a collision with a second liner, the *Algonquin*, was a classic example of the dangers of sailing through busy waters in foggy weather.

The *Fort Victoria* began its life as the *Willochra*, sailing between the United States, Australia, and New Zealand before World War I. During the war it was used as a troopship. The *Willochra* was sold to Furness, Withy and Company of London in 1919, was refitted and renamed, and then placed on the New York to Bermuda route as a passenger cruiser.

The *Fort Victoria*'s date with destiny began when the cruise ship sailed out of New York harbor on December 18, 1929. On board were the master, Captain A.R. Francis, the crew, and just over 200 passengers. Later that day the *Fort Victoria* came to a halt at the beginning of the Ambrose Channel. Fog surrounded the ship and visibility was very poor. Captain Francis could hear warning bell and sirens coming from several directions, but nothing prepared him for the sight of a ship's bow that suddenly appeared out of the fog. The bow belonged to the *Algonquin*, out of Galveston. A collision could not be avoided and the *Algonquin* cut through the *Fort Victoria*'s hull on the port side. It was a death blow and both ships immediately sent out distress calls.

The calls for help were answered rapidly by the US Coast Guard and other vessels in the area. Captain Francis watched as all his passengers and crew were evacuated safely and then left the foundering liner. The *Fort Victoria* sank later the same day.

Left: *The* Fort Victoria *was a cruise ship that worked the New York to Bermuda run in the 1920s.*

Between 1930 and 1939 long-distance passenger travel by air became a viable proposition, at least for the wealthy. Speed was not the issue, rather comfort and style. The vast airships that offered the possibility of such travel were akin to the great luxury liners that plied the world's sea passages. Like the liners, however, the airships were not invulnerable to disaster. Two events, one in France and the other in the United States, undermined public faith in airships. In 1930 the British *R101* succumbed to weather and technical failings, and the mighty German *Hindenberg* followed suit in 1937. These two tragedies, the latter memorably caught on film, heralded the end of the world's brief love affair with airship travel.

Airships were for the rich, but the bulk of people journeyed by train. Railroads were certainly not immune to catastrophe. Thanks to the growth of cities and suburbs, more and more people were being taken to work by train, while holidaying was of increasing importance and the transport of freight was still a top priority. Operating and safety systems could easily be placed under stresses that they were not designed to cope with. When coupled with bad weather or human error, these acute system failures could produce severe accidents. Although rail disasters far outnumbered those involving other forms of transport, public faith in the system was never seriously undermined in the age when steam ruled supreme.

Background: *The twisted remains of the British airship* R101, *which crashed in northern France on October 5, 1930, shortly after the beginning of a flight to India.*

1930–1939

COSTESTI CHURCH, ROMANIA

APRIL 18, 1930

Costesti is a small and remote town in the foothills of the Carpathians about 60 miles (96km) northwest of the Romanian capital, Bucharest. On the evening of Good Friday in 1930 the Orthodox church there was destroyed by a fire in which 140 worshippers died and many more were seriously injured.

The little parish church, a wooden structure more than 100 years old, was packed for the evening service. A lighted candle held by one of the congregation accidentally set fire to a decorative wreath – in a moment the whole building was filled with flames and smoke.

The congregation immediately panicked and rushed straight for the door. But tragically the heavy wooden door opened inwards and the force of the crowd crushing against it kept it tightly closed. Everyone was trapped inside. A few people standing outside in a passageway tried to push their way in, but they were unable to do anything because of the enormous weight of people behind the door.

The alarm was raised immediately but the church was some distance from the town and by the time help arrived it was too late. Someone telephoned the nearest fire station at Pitesti, but this was 15 miles (24 km) to the north and the fire engines drove up just as the roof collapsed. The 140 men, women, and children still inside the church were buried under a heavy mass of burning wood. Two hours later, the fire was extinguished – but all that remained of the church was a smoldering ruin.

Above: A traditional Romanian village church. It was in such a church at Costesti that 140 people perished.

Right: The entrance to the church was a heavy timber door that opened inwards.

ALLONNE, NORTHERN FRANCE

OCTOBER 5, 1930

Much time and considerable amounts of money were invested in an ambitious British airship program in the 1920s. The long distances that airships were capable of flying seemed an ideal means of communicating between the far-flung corners of the British empire.

Two airships, the *R100* and the *R101*, were on the drawing board by 1924. The design specifications for the R101, to be built at the Royal Airship works at Cardington, England, called for a machine capable of flying the 3500 or so miles (7240km) to the Indian sub-continent, with a payload of 30 tons made up of passengers, their baggage, and cargo such as mail. The proposed design was a mammoth 724 feet (220m) long, by 63 feet (19m) high.

During the first six years of development the program ran into many teething problems. The gas bags were prone to rub on the rigid frame and split. Coupled with this, the weight of the giant airship far exceeded the parameters set by the design team, forcing them to undergo an extensive program of weight saving. Among other things, two of the lavatories in the passenger car were removed. To increase the lifting capacity an extra 53 feet (16m) was added to the structure, increasing the length to 777 feet (236m).

By the time the *R101* was anchored at its mast at Cardington for the maiden voyage to India in September 1930 the program was well over schedule and hugely over budget. The ever eager British press, and the Secretary of State for Air, Lord Cardington, continued to herald the airship as a triumph of British aviation. It was certainly an imposing sight, and given the ambitious specifications that the technical staff were given to work with, a triumph over many adversities. But the *R101* was never tested adequately. Just one test had been deemed necessary before the airship was ready for the flight to India. This test flight had been conducted in near-perfect weather conditions.

Fifty-four people boarded the *R101* in the early evening of October 4, 1930, including many high ranking Air Ministry officials and the leading figures in the airship development program. Incredibly, the airship was awarded its Certificate of Airworthiness only minutes before take-off. At 1900 hours the giant airship rose into the air and set a southeast course for Paris, France.

Problems arose almost immediately with one of the engines, which was shut down. At barely 40mph (64km/h) the *R101* flew over London, and three hours after departing from Cardington crossed the English

coast near Hastings. By now the weather had deteriorated considerably and the airship was being buffeted by high winds. As it began drifting eastwards off its course to Paris, engineers realized that rain soaking into the outer skin had also added nearly three tons in weight to the airship.

At just a few minutes past 0200 hours on October 5, as the airship struggled to make headway against a

Above: The burned-out skeleton of the R101 *lying on the side of Beauvais Ridge near the village of Allonne in northern France.*

Above: *Rescue workers remove a body from the scene of the* R101 *crash.*

Left: *After the inferno that consumed the* R101 *airship, little remained but the rigid frame.*

strong headwind, the outer skin on the nose was torn. Soon after, the two forward gas bags were ruptured, and as gas rushed through the gaping tears, the nose of the *R101* pitched down. At this point the watch commander sealed the fate of many of those on board by ordering the engine power to be reduced. This action cancelled out any aerodynamic lift generated by forward motion of the airship, which might have allowed the airship to continue flying.

The *R101* struck the side of the Beauvais Ridge near the village of Allonne at 0209 hours. Fire broke out in the control car directly under the main gas bags, and in seconds the inferno had swamped the whole machine. Four of the engineers and a wireless operator managed to scramble clear.

The loss of the *R101* led directly to the cancellation of the airship program in Britain, although the German nation championed this type of flying machine for another six years.

GEORGES PHILIPPAR, GULF OF ADEN

MAY 15, 1932

The French passenger liner *Georges Philippar* was no stranger to fire. The ship it was built to replace, the *Paul Lacat*, was burned out in Marseille harbor in December 1928 and the *Georges Philippar* itself was ravaged by fire on November 29, 1930, some three weeks before its launch. The interior was badly damaged but fortunately its luxurious fittings were yet to be added. The ship was finally completed in January 1932.

The ship's first and last voyage began under a cloud – French police warned the *Georges Philippar*'s owners, Messageries Maritimes, that threats had been made to destroy the vessel on February 26. The outward journey to Yokohama passed without incident. The ship made a quick turnaround and headed for home, calling first at Shanghai and then Colombo. From Colombo the vessel headed across the Indian Ocean, carrying 518 passengers and 347 crew members. Twice during this leg a fire alarm went off in a store room containing a large quantity of bullion. On both occasions no fire was found.

A fire did break out early in the morning of May 15 in a cabin. The danger was only belatedly reported to the master, Captain Vieg, and by this time the fire had spread. The captain decided to head for Aden at speed and beach the vessel, but unfortunately the high speed only served to fan the flames.

Vieg realized that the fire was out of control and opted to abandon ship. Three vessels answered the *Georges Philippar*'s distress call and rescued over 650 of those on board. Estimates of the fatalities varied between 40 and 90. The *Georges Philippar* burned and drifted for four days before sinking on May 19.

Below: The French liner Georges Philippar was an attractive-looking ship. But its maiden voyage to Yokohama and back was never completed – the ship succumbed to fire in the Gulf of Aden and sank.

PIETER CORNELISZOON HOOFT, AMSTERDAM

NOVEMBER 14, 1932

The life of the Dutch liner *Pieter Corneliszoon Hooft* was destined to begin and end with fire. Commissioned by the Nederland Line from a French shipyard, the ship had its first brush with disaster on December 25, 1925, while it was still in the shipyard. The vessel was engulfed by a major fire that left its passenger accommodation in ruins. The French builders could not meet the deadline for finishing construction, so the ship was sent to Amsterdam for completion.

Amsterdam was to prove a fatal home for the *Pieter Corneliszoon Hooft*. The ship was eventually completed and delivered to its owners in August 1926. Its maiden voyage, from Amsterdam to the Dutch East

Indies, was made the same year. It was a money-spinning route and the Nederland Line decided to improve the *Pieter Corneliszoon Hooft*'s performance. In 1930, the ship was lengthened by nine feet (3m) and was fitted with new diesel engines. Its first voyage after the refit took place in April 1931.

The ship did not survive for long after its refit. On November 14, 1932, the liner was engulfed by fire while docked at Amsterdam's Sumatra Quay. The local emergency services reacted quickly to prevent damage to the harbor, using tugs to tow the *Pieter Corneliszoon Hooft* to safe water, but the fire was out of control and the Amsterdam fire services were unable to save the liner. The vessel was left far beyond repair by the inferno and had to be sold for scrap.

Above: When the Pieter Corneliszoon Hooft *caught fire while berthed in Amsterdam harbor, tugs were quickly mobilized to tow the ship away from the dockside.*

L'ATLANTIQUE, ENGLISH CHANNEL

JANUARY 4, 1933

The Compagnie de Navigation Sud Atlantique was justifiably proud of the 42,512-gross ton *L'Atlantique* – it was the largest and most luxurious passenger ship plying the route to South America. The *L'Atlantique* was launched in 1931 and made its maiden voyage, between Bordeaux and Buenos Aires, on September 29, 1931. The disaster that befell the *L'Atlantique* 15 months later, however, led to a bitter law case and the vessel never put to sea again.

The fateful incident took place as the *L'Atlantique* was heading from its home port of Bordeaux to Le Havre for its regular yearly spell in drydock for maintenance. At about 0330 hours the ship's master, Captain Schoofs, was informed that a fire had broken out on E Deck in cabin 232. The fire spread rapidly and the crew was forced to abandon ship. In the confusion 17 lives were lost. The blaze continued for two days. The *L'Atlantique* drifted toward the coast of southwest England before it was taken in tow by tugs from France, Germany, and the Netherlands on January 6. It was docked at Cherbourg.

When the damage was assessed, the owners claimed the ship was a write-off and put in a claim for its full insured value. The insurers, however, disagreed, estimating that the vessel could be repaired for considerably less. The owners won the subsequent legal battle, and the *L'Atlantique* never sailed again. In 1936 it was sold for its scrap value.

Below: The French liner L'Atlantique *ablaze in the English Channel in January 1933. The ship was eventually taken in tow to Cherbourg.*

REICHSTAG, BERLIN, GERMANY

FEBRUARY 27, 1933

The political repercussions of the fire which largely destroyed the German parliament building were even greater than the conflagration itself. The German legislature was gutted during the night of February 27 by a fast-moving fire. The blaze was started deliberately – probably by someone setting a match to furniture piled on rugs inside the building – but at the time it was not clear who had done it. It is now known to have been the work of the ruling Nazis, who started the fire as a pretext for clamping down on opponents of their rule, particularly the communists, in the build-up to the elections of March 5.

Smoke was first noticed by a police officer on patrol in the Reichstag itself in the evening at about 2100 hours. The officer later claimed that, before raising the alarm, he had fired several shots at a group of men seen running away from the scene. He then managed to arrest one of them, a 24-year-old Dutchman named Marinus van der Lubbe, who – when searched – was found to be carrying a Communist Party membership card.

By the time the fire brigade arrived, the fire was well entrenched and had spread in many directions. The panelling, chairs, and desks of the Reichstag chamber were all made of wood and so burned easily.

The firemen fought bravely and brought the inferno under control before it reached the cupola. They also managed to save the library and reading room where countless priceless documents were stored.

Chancellor Adolf Hitler lost no time in linking the fire to the German left wing and alleged a conspiracy:

Below: A fire engine stationed outside the Reichstag building as it burns on the night of February 27, 1933.

"Now you can see what Germany and Europe have to look for from Communism," he declared. He placed Hermann Göring, later to be head of the Luftwaffe, in charge of the investigation into the fire. Before dawn, police had arrested all 100 elected communist members of the Reichstag. They also rounded up other communists all over Berlin and detained them pending the outcome of the investigation at the scene.

The following evening, President von Hindenberg was persuaded to sign an emergency decree which suspended constitutional guarantees of individual freedom, freedom of the press, private property, and the secrecy of postal communications. Communist newspapers were shut down and suspected communist meeting places were closed.

This was originally supposed to have been a temporary measure, but it remained in force for 12 years: in that sense, the Reichstag fire marked the true beginning of the Nazi dictatorship.

Although Marinus van der Lubbe was mentally handicapped and almost certainly not the culprit, he was subsequently executed.

Above: The fire brigade tackling the blaze inside the Reichstag. It seems probable that the fire was set in several different places.

Right: Firemen surveying the damage to the interior of the Reichstag. They had managed to save the library and the reading room with their collections of priceless documents.

ATLANTIC OCEAN, OFF NEW JERSEY, USA

APRIL 4, 1933

The USS airship *Akron* and her sister ship *Macon* were designed to carry fighter planes in a docking bay and hangar. These planes, it was thought, could both provide protection for the airship and scout for enemy forces. Aircraft were still hampered by limited range and an airship aircraft carrier would provide the ultimate mobile base on the new, modern battlefield.

The *Akron* had less structural bracing than most German airships, and another major difference was that the *Akron* had internally mounted engines. The airship was filled with nonflammable helium, which meant that the power plants could be mounted inter-

nally for ease of maintenance and improved streamlining. The *Akron*'s eight props were built on outriggers on either side of the ship.

On April 3, 1933, the *Akron*, under the command of Commander Frank McCord, was flying off the coast of New Jersey, fighting its way through a thunderstorm. Also on board were Rear Admiral William Moffett, chief of the Navy Bureau of Aeronautics, and Commander Frederick Berry, commander of the Lakehurst Naval Air Station. The storm that the *Akron* was flying through was one of the worst that the New Jersey coast had seen in years. Because of the winds, heavy cloud cover, and ground fog the navigator was unable to determine their position. The captain

Left: The Akron *was a US naval airship designed as an aircraft carrier that could transport fighter planes long distances to where they were needed.*

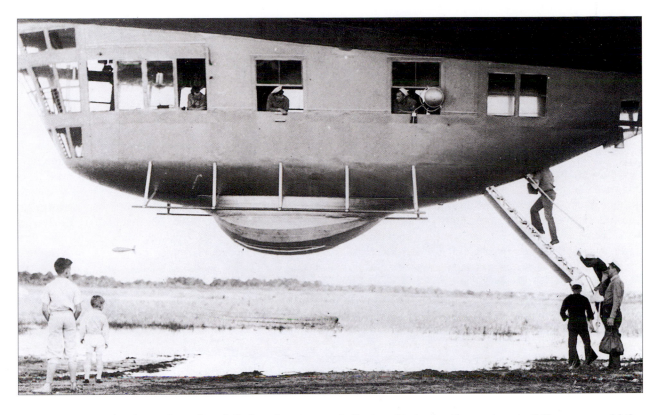

Right: US naval personnel boarding the Akron *at a naval airbase in the year before the disaster.*

decided to turn back to shore but failed to locate a break in the storm. The order was given to head back out to sea and simply ride out the weather. Eventually the *Akron* found some calm air.

At 0015 hours the *Akron* was rocked by a horrific blow. The calm that the unfortunate airship had felt was in fact the eye of the storm system. As it emerged from the other side of the eye, fierce winds caused the massive airship to buck wildly. The ship hit an enormous downdraft and was sucked down almost 1000 feet (300m). Only by dropping most of the water ballast and bringing the engines to full power could the descent be stopped.

The *Akron* was able to gain altitude briefly, but was pushed down again and again until finally the force of the storm started to rip the ship apart. With loss of control of the fins the *Akron* smashed into the freezing water of the Atlantic.

Because the *Akron* carried no life vests and there had not been enough time to lower her one life raft, 72 of her 76-man crew drowned, including Rear Admiral Moffett along with Commanders Berry and McCord.

Below: The wreck of the Akron *adrift in the storm-tossed seas off New Jersey.*

LAGNY-POMPONNE, FRANCE

DECEMBER 23, 1933

This rear-end collision led to the death of 230 passengers and left 300 injured. The incident began with the late running of an express traveling from Paris to Nancy. It was delayed because of thick fog and was brought to a halt by adverse signals at Pomponne, a short distance outside Lagny. As it was moving off, it was hit by a second express, one traveling from Paris to Strasbourg, at 60mph (96km/h). The second train had also been delayed because of the fog and frost.

The Paris–Strasbourg express smashed its way along much of the Nancy train, which was packed with Christmas shoppers, and five of its carriages were battered beyond recognition. The vast majority of the casualties occurred in the Nancy express as the steel carriages of the Strasbourg train stood up to the impact remarkably well. Many of the injured were ferried back to Paris in those coaches that had not been derailed by the initial collision.

Charges were brought against those held responsible, including the driver of the Paris–Strasbourg express, and the trial began in December 1934. He was acquitted and the cases against other employees were also dropped when the court delivered its verdict in January 1935. It was acknowledged that he had been traveling far too fast given the foggy and frosty conditions, but that he may have missed stop and warning signals either because of the fog or because the signal mechanisms had been adversely affected by the frost.

Below: The crushed remains of the locomotive that hauled the Paris–Strasbourg express, one of the two trains wrecked in the crash at Lagny-Pomponne.

DRESDEN, NORTH SEA
JUNE 20, 1934

The sinking of the *Dresden* while on a pleasure cruise illustrates how quickly a carefree outing can turn into a time of terror. Many of the ship's young passengers could not swim, and were drowned when the ship struck a rock.

The German passenger liner *Dresden* had a varied career. The ship was built in 1915 and named the *Zeppelin*, but it spent the remainder of World War I laid up. Germany's defeat saw the *Zeppelin* handed over to Britain as part of the war reparations. The vessel was sold on to the Orient Line in 1920, refitted, and renamed the *Ormuz*. Under this name the liner was put to work on the London to Australia route.

The *Ormuz*'s career sailing to Australia ended in April 1927, when an offer for the ship by the North German Lloyd Line was accepted. The liner underwent another refit, was renamed the *Dresden*, and started operating between Bremerhaven and New York. However, part of the *Dresden*'s time was spent as a cruise ship, taking poorer Germans on trips on behalf of the Nazi Party's "Strength Through Joy" campaign.

The *Dresden* was on one such cruise in 1934 when the disaster happened. At about 1930 hours on June 20 the liner struck a submerged rock off Karmoe Island. The impact reverberated around the ship and the passengers, many of whom had never been to sea before and could not swim, began to panic. Some threw themselves into the water. The ship was beached eventually, but with water pouring in through three holes in its hull, there was no chance of it surviving. The *Dresden* began to list and keeled over on to its side the next day. The passengers and crew who survived were taken to several Norwegian coastal towns.

Above: The capsized Dresden. *The ship was taking a party of young Germans on a pleasure cruise when it struck a rock.*

MORRO CASTLE, ATLANTIC

SEPTEMBER 8, 1934

The fate of the *Morro Castle* was a classic example of how incompetence and negligence can result in unnecessary loss of life. When the *Morro Castle* caught fire it was only six miles off the coast of New Jersey. Yet the confusion that reigned on board led to the loss of 137 lives – and a series of expensive law suits for payment of damages. Another result was that new regulations governing safety at sea were brought in for US ships.

The *Morro Castle* was built to transport passengers between New York and Havana and undertook its maiden voyage out of New York on August 23, 1930.

Some four years later, in September 1934, the *Morro Castle* was nearing the end of the homeward voyage with 316 passengers and 231 crew members on board.

The captain, Captain Robert Wilmott, appears to have been a rather paranoid character. He was reportedly convinced that someone was out to do him harm and so isolated himself, staying either on the bridge or in his cabin.

Wilmott never made it back to New York alive – he died from a suspected heart attack during the night of September 7 and was replaced by Chief Officer William Warms. When Warms took over command of the liner, it was negotiating heavy seas and strong winds. At

Left: When the acting captain gave the order to abandon ship, many of the first people into the lifeboats were actually members of the crew.

Above: The still-smoldering Morro Castle *lies beached in front of Convention Hall, Asbury Park, New Jersey.*

Right: A coast guard returns from the charred remains of the beached Morro Castle *carrying the body of a child found dead on the wreck.*

about 0200 hours on the morning of September 8 a passenger found a fire in a writing room. He called a steward, who attempted to put out the fire. Unfortunately the steward failed to notify the bridge about the problem. About 60 minutes passed before Warms heard of the fire, by which time it had spread

alarmingly due to the ship's wooden panelling and the strong winds that were blowing. To make matters worse, the ship was short of the correct number of firehoses, and those that were available had to be dragged some distance to fight the fires. Many of the water points had been capped, making them unusable.

There had been no fire or boat drill during the voyage, so neither the passengers nor the crew members knew what to do in the emergency and confusion reigned. Warms was too inexperienced to follow correct procedures. He did not send out a distress call for some time, even though the fire was clearly getting out of hand. Distress signals were sent out eventually but these ceased when an explosion destroyed the generator.

The crew behaved badly once the evacuation got underway. Passengers and crew gathered at either the bow or stern as the central section of the ship was ablaze. The first lifeboat to get way held 92 crew members and only six passengers. Several vessels picked up the *Morro Castle*'s distress call and raced to the scene. These ships rescued many of the passengers, but 137 people died. The abandoned liner eventually drifted ashore at Asbury Park, New Jersey.

The subsequent board of investigation placed the blame for the catastrophe on the shoulders of the ship's owners and crew. Warms was found guilty of negligence and given two years in prison, the vessel's chief engineer, one of the first into a lifeboat, was jailed for five years, and the vice-president of the line was fined. Passengers received compensation.

GROSS-HERRINGEN, GERMANY
DECEMBER 24, 1935

Above: The scene at the bridge over the River Saale where the crash occurred. Some wreckage is strewn along the foot of the embankment and rescuers are clearing the track of obstructions.

This Christmas Eve incident involved a local train traveling from Erfurt to Leipzig and the two-engined Berlin–Basel express. Over 30 passengers were killed and 27 seriously injured because of the failure of the driver of the front locomotive of the express to heed two sets of warning signals. Casualties might have been even higher, but for the fact that one of the drivers of the express was able to slow his locomotive somewhat after he had noticed the second warning signal.

As the local train was leaving the junction at Gross-Herringen station and was crossing the main line on a bridge over the River Saale, it was struck on the side by the speeding express. Several of the carriages of the Erfurt–Leipzig train were crushed, one was thrown on top of one of the express's locomotives, and a goods van was left hanging off the bridge.

None of the local train's carriages fell into the River Saale, although initial reports suggested that a large piece of wreckage found in the river by emergency service workers was in fact a carriage. However, several bodies were later recovered from the icy waters and several more were reported missing so the final death toll may have been higher than the official statements of the time suggested. Passengers in the express received comparatively minor injuries as they were offered a measure of protection by the train's all-steel carriages, which stood up to the violent impact surprisingly well.

SHRIVENHAM, ENGLAND

JANUARY 15, 1936

This collision on the Great Western Railway involved a coal train traveling from Aberdare to Old Oak Common and a nine-carriage sleeper making its way from Penzance in Cornwall to London's Paddington station. A collision between the two left two people dead, and the chief investigator, Lieutenant-Colonel Mount, concluded that errors committed by two signalmen were the cause of the smash.

The incident began when the coal train pulling 53 full wagons broke in two a little time after passing through Swindon station. It had been a bitterly cold night, with frost and low-lying mist, and dawn had yet to break. Five wagons and the brake van were detached from the rest of the train when a draw hook failed and came to a stop a short distance outside Shrivenham station in Wiltshire.

Unfortunately, neither the signalman at the Shrivenham box nor a colleague located in the box at Ashbury Crossing, a little way down the track to the east, noticed the dangerous blockage on the line. Both also failed to note that the coal train had passed and did not have any rear warning lights. These were on the stationary brake van.

Both gave the all-clear to the sleeper express from Penzance, which was following closely behind the coal train. However, disaster could still have been avoided if the guard on the coal train had gone back down the track waving his red light to indicate the blockage. He

Above: One of the carriages of the sleeper express wrecked in the collision at Shrivenham, Wiltshire.

Above: The aftermath of the incident at Shrivenham. Here, police and rail officials examine the remains of the Penzance sleeper express.

did not do so, and the high-speed impact became inevitable. Shortly before 0530 hours, the sleeper ran into the halted brake van and coal wagons at a speed of approximately 50mph (80km/h), killing the driver of the express and one of its passengers. It was a remark-ably small number of fatalities given the speed at which the crash took place.

Both the express's driver and fireman had spotted the three red warning lights on the back and sides of the coal train's detached and stationary brake van and the driver applied his brakes, but was unable to stop before hitting the obstruction.

The brake van and three of the loaded coal wagons were smashed in the impact and the two other wagons were propelled some considerable distance farther down the track. The locomotive pulling the sleeper,

"King William III," rode over the wreckage and was sent crashing on to its righthand side. Its lead carriage, containing 34 passengers, suffered varying degrees of damage.

The front half was comparatively unscathed, but five of its rear compartments were destroyed. Most of the 10 people severely injured in the collision were found in this section, and the only passenger fatality, a woman, was also discovered here.

The second carriage of the express was also completely destroyed, but luckily it was empty at the time of the crash. The driver of the express locomotive, E. Starr, was badly injured in the collision and rescuers took several hours to free him, but he later died from his injuries. The fireman of the sleeper escaped with relatively minor injuries.

CRYSTAL PALACE, LONDON, ENGLAND

NOVEMBER 30, 1936

Crystal Palace – a huge glass conservatory built to house the Great Exhibition of 1851 – was originally sited in Hyde Park. After the exhibition it was moved to the top of a ridge 300 feet (90m) high at Sydenham in southeast London, where it was turned into a concert hall and the focal point of a great amusement park. Here it remained until the last night of November, 1930, when it was razed to the ground in a fire so spectacular that it was clearly visible from as far away as the coastal towns of Brighton and Margate.

Flames were first noticed in the central part of the great structure at about 1930 hours. A musician who had just arrived to take part in a choir practice saw the fire but was told not to worry, it would soon be out, and to carry on with his rehearsal. Three hours later, however, more than two-thirds of the Crystal Palace were lying on the ground in a flaring mass of ruins. Even then the fire raged on unchecked, fanned by a strong northwesterly breeze.

The roof – 400 tons of glass – had been the first part of the structure to succumb and crash to the ground. Then the huge skeleton of the building – 4000 tons of iron girders – bent and twisted before falling in an immense shower of sparks. The Palace's 200 miles (300 km) of wooden window frames burned like match sticks and helped to spread the conflagration.

When the alarm was raised the permanent firefighting force attached to the Palace did their utmost with the limited resources at their disposal, but it quickly became apparent that their efforts were hopeless. By the time the first outside aid arrived from the fire station at nearby Penge, the whole of the central transept was ablaze. Before long, firemen from all over London arrived to help – their 20 hoses needed so much water that domestic supplies were reduced to half pressure throughout the night. Even this was insufficient, and the emergency reservoir and tanks had to be brought into use.

Flames rose more than 150 feet (45m) above the Palace. The pilot of an Imperial Airways aircraft reported that he could see the fire from the middle of the English Channel. Thousands of people throughout

Left: The Crystal Palace was built by Joseph Paxton to house the Great Exhibition of 1851 in Hyde Park. The building was later dismantled and moved to Sydenham.

London and southeast England noticed the great glare in the distance and set out to get a closer view of the inferno – the destruction of this great building became a tourist attraction, later described as "the best show in town." Special trains were run to bring sightseers from central London and there were mile-long traffic jams on all roads leading to the fire. Yet, although there is no doubt that these crowds hampered efforts to contain the fire, their interest was not altogether ghoulish – throughout the throng there were expressions of sadness at the end of "the poor old Palace."

Despite the size of the blaze, no one was seriously injured. The main casualties were thousands of birds which were released from an aviary in the grounds for their own safety but were then overcome by smoke and crashed to their deaths in the flames.

Above: After the fire little remained of the great glass and iron Palace but a mass of black, twisted girders.

Left: The death of an historic building. The heat of the fire melted the iron structure and brought the glass crashing to the ground.

LAKEHURST, NEW JERSEY, USA

MAY 6, 1937

There are few more arresting images of the 20th century than that of the Zeppelin *Hindenburg*, its aft quarters engulfed in a ball of flame, lighting the night sky above Lakehurst, New Jersey. The *Hindenburg* was the largest of the many rigid airships – dirigibles – built in Germany between 1900 and 1940, and is still the largest such craft ever constructed. Its loss was a great blow to the prestige of the German airship industry and its manufacturer, Luftschiffbau Zeppelin (LZ). Confidence in the airship was never fully restored after the disaster.

The *Hindenburg* was the last but one in the LZ series (*Luftschiff* means airship), and a direct descendant of the Zeppelins that had struck awe and terror into the hearts of Londoners during World War I. The airship had proved too vulnerable as an offensive bombing platform during that conflict, and so the numerous airship manufacturers in Europe and America had begun to look for alternative applications for the technology.

Luftschiffbau Zeppelin, the company established by Ferdinand Graf von Zeppelin in 1900, seemed the most likely to succeed. The 100-hour flights that were a regular feature of the Zeppelin raids on England made

Below: The Hindenburg sails over New York, en route to Lakehurst, New Jersey, after a transatlantic flight from Germany in 1936. On the left is the Empire State Building.

Above: The newly-completed Hindenburg *emerging from its shed for the first time.*

circumnavigation of the globe in 21 days. When the *Graf Zeppelin* was finally decommissioned in early 1937, the future of the airship seemed bright.

The *Hindenburg*, the airship built to supersede the *Graf Zeppelin*, was a truly awe-inspiring sight. Designed by Dr Hugo Eckener, the foremost airship expert of the day, the airship measured 804 feet (245m) from tip to stern. It was powered by four 1,100hp diesel engines located in streamlined nacelles attached at points on the side of the light, but strong, duraluminum rigid frame. The gas cells that provided the lifting power were filled with volatile hydrogen – a lighter-than-air gas.

Eckener's original intention was to fill the cells with nonflammable helium. Unfortunately for the Germans about 90 percent of the world's supplies of helium at this time were produced in a small area of Texas. With an airship program of is own the US Congress was unwilling to allow the expensive gas to be exported.

The presence of huge quantities of a highly inflammable gas, some have argued, made the *Hindenburg* little more than an accident waiting to happen – a flying bomb. The design team were certainly aware of the dangers; the gas bags were brushed with a gelatin solution to prevent gas escaping, and both passengers and crew had to observe rigorous safety procedures. The men patrolling the walkways wore hemp-soled shoes (to prevent any chance sparks from boot nails) and anti-static overalls. In the smoking room, a steward stood watch night and day and the walls were

transatlantic flight a realistic possibility. Such performances, far in excess of anything attainable by conventional fixed-wing aircraft at the time, encouraged the company to experiment with commercial passenger-carrying airships, and in September 1928 LZ inaugurated a transatlantic service with the *Graf Zeppelin*. This airship completed 590 flights, 144 of them across the North Atlantic, for a total of more than 1,000,000 miles (1,600,000km). In 1929 the airship made a

Right: The last moments of the Hindenburg, *as it bursts into flames close to its mooring mast at Lakehurst airport, New Jersey.*

Right: The Hindenburg was filled with highly combustible hydrogen gas – so the slightest spark would turn it into a fireball, as did indeed happen at Lakehurst.

lined with pigskin to further prevent the risk of fire. Passengers and their luggage were searched for potentially incendiary materials before boarding.

The interior of the airship was splendidly fitted out in the fashion of the luxurious ocean liners of the day – which many thought the airship would eventually replace. Sumptuous velvet upholstery and walnut coachwork graced the passenger suites and dining salon, and in the fully-equipped kitchen, staff prepared elaborate meals for the complement of up to 50 well-heeled passengers. There was even a grand piano, albeit one fashioned from duraluminum and weighing only 112lb (50kg). The airship was the Concorde of its day – a luxury transatlantic ferry and a potent symbol of the technological advancement of Germany under the Third Reich.

The *Hindenburg* made its maiden flight in April 1936, and during the course of the following 12 months the airship made 10 successful and highly publicized transatlantic crossings.

At 2000 hours on May 3, 1937, Captain Max Pruss, the veteran of 16 transatlantic crossings in the *Graf Zeppelin*, gave the order "Up ship," and the *Hindenburg* rose into the air for the first flight to America of the year. At sunrise on May 4 the Hindenburg was passing south of Ireland, heading toward the coast of Newfoundland. The following day the airship ran into headwinds and Pruss was forced to radio ahead to Lakehurst that he expected his landing to be delayed by 12 hours, until 1800 hours on May 6.

The airship flew over Boston and descended to 600 feet (180m) to give the admiring crowd below a good view of the pride of the Zeppelin company. The airship then proceeded down the coast and over the New York skyline. Pruss delayed his approach to the mooring mast at Lakehurst until 1900 hours because of reports of bad weather, but at 1900 the crew were finally given the order to prepare for landing. Gas was vented off, and the landing wheels were released.

At 1921 hours, the lines that would be used to winch the craft down had just been released, when an explosion ripped through the No. 4 gas cell just in front of the fin. In seconds the whole of the after part of the ship was engulfed in a huge fireball, and burning wreckage began falling on the ground crew below. The *Hindenburg* lurched and the tail began to sink to the ground. Passengers leaped for their lives from the control gondola and the crew quarters, and in only 30 seconds the giant airship lay in a smoking ruin on the ground. Remarkably, 62 out of the 97 people on board escaped with their lives, including 23 of the 36 passengers.

Many explanations were forthcoming during the months of investigation by the US Navy and the German government. One suggestion was that the explosion was caused by a device planted by one of the riggers as a demonstration against the Nazi government. A rather more likely explanation was that gas which had escaped from a ruptured gas bag was ignited by the static electricity that had built up in the air during that stormy New York night.

No final conclusion was ever reached as to what had caused the explosion that destroyed the *Hindenburg*. One thing was certain, however. The disaster was a death blow for the airship industry.

PARIS, LE HAVRE
APRIL 19, 1939

The owners of the *Paris*, the Compagnie Générale Transatlantique, wanted to build a palace of the high seas to cash in on the highly profitable transatlantic passenger service. Its first-class passenger rooms were sumptuously decorated in the art nouveau style, and the whole ship was designed to give passengers the feeling that they were already in France.

Below: After being ravaged by fire all night, the French liner Paris *sinks in the dock at Le Havre.*

Work began on the liner shortly before World War I. Construction of the *Paris* was started by Chantiers et Ateliers de St Nazaire in 1913, but when war broke out work was stopped in 1914. With slipway space in high demand, the building was subsequently resumed and

the *Paris* was launched in mid-September 1916. When the fitting-out work was completed in June 1921 the *Paris* was, at 34,569 gross tons, the largest passenger ship ever built by French yards. Its owners had also ensured that its more than 550 first-class passengers would enjoy luxurious surroundings and almost unparalleled on-board facilities. The *Paris* made its first transatlantic trip from Le Havre via Plymouth to New York on June 15, 1921.

The first disaster to strike the *Paris* occurred in August 1929 while the vessel was docked at Le Havre. A fire broke out, which destroyed a large part of the ship's passenger accommodation. It is possible that the fire was due to arson, since the shipping company had

received a warning that its ships might be sabotaged. Faced with an extensive refit, the company decided to revamp and improve the ship's accommodation. By January 1930 the refitting had been completed and the *Paris* was put back into service.

World War II was just a few months away when fire struck the *Paris* for the second time. The ship was again docked at Le Havre. On April 19, 1939, several fires broke out at the same time – one began in the vessel's bakery and two others started on two of its upper decks. Despite the fact that the ship was in harbor, the emergency services were unable to prevent the fires from spreading and the blaze was soon out of

control. Eventually the *Paris* capsized and sank at the dockside. Only part of its hull and superstructure remained visible above the harbor's waters. Again there were rumors that the *Paris* had been the victim of a deliberate act of arson.

World War II prevented the ship being raised. The last chapter in the sorry saga of the *Paris* began shortly after the end of the war. In 1946 another ship, the *Liberté*, broke free from its moorings and crashed into the remains of the *Paris*. It was the final blow. The *Liberté* was towed free, but the *Paris* was clearly beyond hope of salvation. In 1947 any plans to raise the liner were abandoned and the ship was scrapped.

Right: The Paris *lies on its side in the dock, where it was to remain for eight years before it was finally scrapped.*

The period from 1940 to 1949 was dominated by World War II and its immediate aftermath. The conflict produced some 15 million military losses and more than 30 million civilian casualties. Inevitably, the need to prosecute a war often overrides the absolute safety considerations that would be considered paramount during peace, and modern war itself, as the figures above show, can produce truly enormous civilian casualties.

Clearly, some disasters can be the product of deliberate military activity, not least in the case of the mass bombing of Hamburg (July 1943), Dresden (February 1945), and Tokyo (March 1945), which together produced an estimated total casualty list of between 300,000 and 400,000 people. These events are obviously staggering in terms of the scale of destruction and loss of life caused, but not all disasters were a direct result of the fighting. Fires continued to break out for the usual reasons. For example, Boston's Cocoanut Grove Night Club was ravaged by fire in November 1942 mainly due to a carelessly wielded match. Trains also continued to leave the rails regularly, as did the "Congressional Limited" outside Philadelphia in September 1943, due to technical failure. The end of World War II did not of course signal an end to disasters. Civilians may have been dying in fewer numbers but there were still numerous fatal incidents reported across the globe on a regular basis.

Background: German civilians attempt to go about their normal activities amid the ruins of Dresden, the target of British bombers during February 1945.

1940–1949

COCOANUT GROVE CLUB, BOSTON, USA

NOVEMBER 28, 1942

At the height of World War II, the USA was rocked by a domestic tragedy unconnected with the hostilities. A sudden fire ripped through Cocoanut Grove Night Club in Boston, Massachusetts, at about 2200 hours on the night of November 28, 1942. More than 300 people were killed – some estimates put the figure as high as 474 – and a further 150 were seriously injured.

After extensive interior refurbishment during the previous year, Cocoanut Grove had become one of the most fashionable spots in the Boston midtown district. On the night of the tragedy, it was full to bursting with US sailors, marines, and coastguardsmen – many on leave from active service in Europe – with their wives and girlfriends. There were also many American football fans who earlier that day had been to a game in which under-dogs Holy Cross had beaten red-hot favorites Boston by 55 to 12. The exact number in the club at the time is unknown, but it is thought to have been in the region of 1000.

The new decor featured simulated leather chairs, plastic palm trees, and a silk-draped roof. The style was one of the club's chief draws – but it also made it a deathtrap. At 2217 hours, the performers had just gone out on stage: bandleader Micky Alpert started

Right: The scene outside the Cocoanut Grove club as the dead are carried out in the aftermath of the horrific fire that killed over 300 people.

one of the plastic palm trees, which went up in a moment. The fire spread rapidly and poisonous fumes belched out of the molten plastic walls.

The clubbers panicked and rushed for the doors, but the exits were not signposted and at least one of the escape routes was locked. Those who found their way back to the entrance in the Melody Lounge became wedged in and around the revolving door – many of the victims were found in this area, either crushed or overcome by smoke.

The gallant efforts of Marshall Cook, a south Boston boy dancer, saved at least 35 – mainly artistes – as he led them to safety up a ladder and across the roof to adjacent buildings. Although the fire brigade arrived quickly, they could not get into the club because of the pile of bodies already obstructing the entrance. On arrival at Boston City Hospital, many of the dead were found to be so badly burned that it proved impossible to identify them.

The official inquiry into Cocoanut Grove, published in November 1943, found that the club had been a disaster looking for an opportunity to happen. The decorations were unsafe, the sprinkler system was inadequate, and the exits not properly signposted. The Boston City Building Commissioner and a senior policeman were indicted but subsequently acquitted. However, one of the club's owners was jailed for 12 years for failing to make adequate safety provisions.

Above: Firemen inspect the revolving doors at the rear of the Cocoanut Grove. These led to a tiny vestibule where panicking guests were crushed and smothered as they fought to get out of the burning building.

playing "The Star-Spangled Banner" to announce the beginning of the entertainment. At the same time, a young waiter named Stephen Tomaszewski went to change a broken light bulb on one of the tables.

The club was dimly lit, so he struck a match to help him see what he was doing and accidentally set fire to

Right: A view of the interior of the club after the fire. The plastic palm trees and silk hangings of the club's decor had ignited quickly, making the place a deathtrap.

HAMBURG, GERMANY
JULY 25–28, 1943

At least 45,000 people were killed during a series of relentless attacks on Hamburg by bombers of the British RAF and the US Army Air Force (USAAF). Over 10,000 tons of explosives were dropped on Germany's second city, flattening an enormous area of it. Although strategically important targets such as factories, shipyards, and the tunnel under the River Elbe were destroyed, it was also part of the Allies' intention to terrorize the civilian population by creating firestorms.

A firestorm occurs when a fire becomes so intense that it uses up all the available oxygen in the air around it; then, as hot air rises, fresh oxygen is drawn down from above and replenishes the flames. This rapid movement of air also serves to fan the existing fires and temperatures may reach about 1800°F (1000°C). A firestorm is a terrifying phenomenon, and is virtually unstoppable by conventional firefighting forces.

Because of the intention to start a firestorm, the offensive was code-named "Operation Gomorrah," after the city destroyed by flame in the Old Testament.

The first raid of "Operation Gomorrah" began at about 0100 hours in the morning of Saturday, July 25, and was carried out by 791 RAF Lancaster, Stirling, Halifax, and Wellington bombers which approached central Hamburg from the east, flying in streams between the Alster Lakes and the Elbe. One of the areas worst affected by this attack was Billwerder near the railroad marshalling yards – the fires that broke out there spread quickly north to the Hamm and Borgfelde residential neighborhoods.

Left: The devastation left by the bombing was appalling. Nevertheless, the citizens of Hamburg continued their daily lives as best they could.

Fire also consumed the densely populated old Altona, Hoheluft, and Eimsbüttel districts, where houses began to collapse and block the narrow streets. Soon St Pauli and the waterfront were also alight. Civilian casualties were particularly heavy, not only because of the intensity of the bombing, but also because the British aircraft were dropping newly developed incendiary bombs that were made of phosphorus. These created such an intense heat that – in the words of one eyewitness – "burning asphalt made the streets look like rivers of fire."

Every available fire engine was summoned to help, and by 0410 hours the city had been officially declared a major disaster area. Although strategic targets were hit, the bombing was indiscriminate and when the fires took hold they rampaged through everything in their path, including thousands of private houses.

At daybreak on July 25, a large greyish cloud hung like a pall over the city of Hamburg. Then, at 1634 hours, more than 200 daylight bombers of the USAAF arrived to carry on where the RAF had left off. After that assault, the night was quiet, but on the following day – July 26 – the American Flying Fortresses returned and strafed the stricken city yet again. Even then, the ordeal of the citizens of Hamburg was not over. On the night of July 27, 787 RAF bombers approached the city from the northeast. Nine districts east of the Alster were badly hit, the worst affected again being Borgfelde and Hamm.

By 0200 on the morning of the 28th, the last bombers turned back to base. Behind them, four square miles (6 sq km) of Hamburg were alight and 16,000 buildings were on fire. The firestorm that started on the night of July 27 raged until 0900 hours the following morning. The death toll of the raids and fires was put at 45,000.

Above: A building still burns as people pick their way through the rubble that was once the city of Hamburg.

Right: Some districts of the city were almost entirely flattened by the raids.

FRANKFORD JUNCTION, PHILADELPHIA, USA

SEPTEMBER 6, 1943

Above: A body is recovered from the wreckage of the "Congressional Limited" following the incident in northeast Philadelphia.

This catastrophe began in the leading dining car of the "Congressional Limited" and was instigated by intense heat which led to the shearing-off of part of an axle in contact with a bearing. The electric-powered express was pulling 16 carriages on this particular journey and it was packed with commuters enjoying the comforts on offer on board one of the railroad's most prestigious trains. The mechanical failure could not have happened at a more inopportune moment. The express, one of the most renowned working between Washington and New York, was just picking up speed after passing the slower section of track around Philadelphia. Its published traveling time between Washington and New York was based on an average speed of nearly 65mph (104km/h).

The fatal accident might still have been avoided, however. Some two miles (3.2km) from where the accident did occur, members of the crew of a locomotive shunting rolling stock in a siding spotted the burning axle and reported what they saw to a signal

Above: One of the train's coaches is pulled clear of the main line while rescuers look on. The violence of the impact with the gantry can be gauged by the extensive damage shown.

box. Their desperate warning was just too late to save the "Congressional Limited," as the axle then broke before safety measures to bring the train to a halt and evacuate the passengers could be implemented.

The sheared axle caused the dining car to derail and then collide with a trackside signal gantry that, because of the express's rising speed, sliced through the entire length of the packed dining car, leaving it mangled almost beyond recognition. The next carriage along was also wrecked by the gantry impact and several more were derailed. Rescue workers sifting through the smashed and derailed carriages found 79 passengers dead and more than 100 injured.

Concern was expressed over the mechanics of the dining car, which had been added to the train at the last minute to cope with the rush of weekend passengers, that began the disaster. It was one of only two carriages hauled by the express that day that did not

have a more advanced axle and bearing system. However, much of this speculation was rejected by the crash investigators. They stated that one of the more advanced coaches had experienced a similar, if less fatal, accident only a few months prior to the Frankford Junction incident and that the high death toll on the "Congressional Limited" was not caused by the derailment itself but by the unfortunate collision with a substantial piece of trackside equipment – the signal gantry. Such an object had not been present in the derailment of the more modern carriage.

Commentators also noted that much of the criticism directed at the dining car was generated by individuals who had items of modern rail equipment to sell to the various railroad companies and that, consequently, their judgment may have been clouded by questions of economics rather than being based on issues of passenger safety.

Right: The signal gantry that sliced through the train's dining car causing many deaths and injuries was also badly damaged by the high-speed collision.

RENNERT, NORTH CAROLINA, USA

DECEMBER 16, 1943

This collision, which left 74 dead and 54 injured, began when the southbound "Tamiami Champion" shuddered to an unexpected halt. Two of its crew descended to the track and began to search for a cause. Between the train's second and third carriages, they spotted a disconnected brake hose and a broken coupling. As one of the crew initiated repairs, the second informed the train's driver to implement safety precautions to protect traffic on the northbound line. Unfortunately, this work was never completed. More importantly, however, no one had immediately spotted the real reason for the train's halt: the last three carriages had been derailed by a faulty rail.

Below: A large crane pulls apart two carriages involved in the accident at Rennert. Many of the casualties were US servicemen.

Two of these carriages, one of which was leaning over at an angle of 45 degrees, were partially blocking the northbound track. Their passengers, none with serious injuries, were evacuated. A crew member then informed the crew stationed at the front of the train – a distance of 450 yards (415m) – of the situation, but failed to ensure that they knew of the derailment.

Some 40 minutes after the derailment, the northbound "Tamiami Champion" reached Rennert traveling at a speed of 85mph (136km/h). Its crew were unaware of the blockage and received a warning only moments before the impact. The collision left the northbound diesel and eight coaches derailed. All but one of the dead were traveling on the northbound train.

WILHELM GUSTLOFF, BALTIC SEA

JANUARY 30, 1945

The 25,484-ton passenger liner *Wilhelm Gustloff* was built by Blohm and Voss, Hamburg, and launched in 1938. The ship had been commissioned by the German Nazi Party's "Strength Through Joy" program, and it was named after a Swiss Nazi leader who had been assassinated in 1936. Intended to give German workers low-cost cruises, it was the first ship built especially for this program, and for this reason it had only one class of accommodation.

Before the *Wilhelm Gustloff* could start on its intended role as a vacation cruise ship, World War II broke out, and the ship was assigned to the German Navy to serve as a troop ship and hospital ship in the Baltic Sea. At the beginning of 1945, the *Wilhelm Gustloff* was acting – along with several other ships – as a rescue ship in the massive evacuation of troops and refugees from the Baltic ports beseiged by the advancing Red Army.

Above: The Wilhelm Gustloff *was launched on May 5, 1938, in the presence of Adolf Hitler.*

Left: Although built as a cruise ship, the Wilhelm Gustloff *was never used in that role.*

At noon on January 30 the *Wilhelm Gustloff* sailed out of the port of Gdynia, Poland, crammed with about 6000 refugees and wounded servicemen. Just after 2100 hours the ship was struck by three torpedoes from a Soviet submarine, and sank almost immediately. Only about 500 people survived – the exact death toll is not known, but it must have been around 5500, making this the worst loss of life recorded in maritime history.

DRESDEN, GERMANY

FEBRUARY 13–17, 1945

Above: One year following the bombings and subsequent terrible firestorm, Dresden remained a wrecked city.

Before World War II, Dresden was known as "the flower of the Elbe" and was numbered among the world's most beautiful cities. The city's prewar population was about 600,000, but this rose during the war to about a million because it was widely believed the Allies would not bomb its historic buildings. During the latter part of the war, however, the city was almost completely destroyed in massive bombing raids carried out by the British RAF and the US Air Force. Much of the damage was caused when repeated bombing created a firestorm that swept through the city.

On February 13 at 2215 hours, 245 RAF Lancaster bombers flew over Dresden and dropped incendiaries and two-ton bombs across the greater part of the city. The attack went on in waves. The incendiaries started thousands of small fires that rapidly came together in a single, great firestorm.

The intense heat of the resulting inferno created a huge column of smoke and flame that rose miles into the air and extended over a wide area. In the center of the vortex there was a terrific updraft of air. This created an area of low pressure at its base, and into the void rushed cooler surrounding air that fanned the flames even further. The winds thus generated were greater than those of a tornado. A tornado is thought to be caused by an increase in temperature of about 20 or 30 degrees Centigrade, but in the Dresden firestorm, the temperature is thought to have risen by as much as 1000 degrees Centigrade. This resulted in a fiery wind that rushed upwards at speeds of more than 100mph (160km/h).

At noon the following day, 450 B-17s of the United States Air Force came across and bombed central Dresden again. Then in the evening the RAF bombers returned – this time there were 550 of them.

German anti-aircraft defenses in the area were inadequate and the Luftwaffe put up very little resistance in the air: only eight Allied bombers were shot down throughout the whole operation. It is impossible to be certain how many people died in these raids: estimates range between 60,000 and 135,000, most of whom were civilians. Total casualties are thought to have been in the region of 400,000.

Dresden continued to be bombarded by raids until April 17. Wave after wave of attacks obliterated the greater part of the city, which was so badly damaged that, at the end of the war later that year, it was suggested that it might be best to level the site and rebuild entirely.

Eventually, however, a compromise was reached: it was decided to rebuild the most important historic buildings, notably the Zwinger – a rococo-style museum on the southern bank of the River Elbe – and the baroque buildings around the castle, and create a new city in the area outside. Much of Dresden was reconstructed with modern, plain buildings, broad streets and squares, and green open spaces, with the aim of preserving as much of the city's former character as was possible in the circumstances.

Above: Dresden in 1946 – piles of rubble are testimony to the city's devastation.

Right: A photograph taken in April 1945 shows the ruins of the museum of fine arts.

TOKYO, JAPAN

MARCH 9–10, 1945

Right: Tokyo under attack. On the night of March 9, 1945, 280 USAAF bombers subjected the capital of Japan to a raid of unprecedented ferocity.

At least 80,000 and possibly as many as 200,000 citizens of Tokyo were killed in a raid of low-level firebombing carried out by the US Army Air Force (USAAF) on the night of March 9–10, 1945. Although the exact number of dead will never be known, it is now thought that this raid claimed more victims than the atomic bomb that was dropped later that year on Hiroshima.

The attack began under cover of darkness at 2200 hours on the night of March 9. It was carried out by 280 USAAF B-29 bombers which had been specially converted for the operation: all their armaments except their tail guns had been removed to accommodate the maximum weight of incendiary devices.

The first wave came in at an altitude of 5000 feet (1500m) and dropped napalm markers in the shape of a cross over the southeast of the Japanese capital. Subsequent streams of aircraft returned to describe a flaming circle around the four points of the cross. After

that, the object was to set fire to all four dark quadrants – an area of 15 square miles (40 sq km).

The night was clear and there was a strong breeze gusting to 40mph (65km/h). The wind fed the fire and the fire fed the wind until the blaze intensified into a great firestorm. Flames shot up in great dragon's tails so high into the air that the undersides of the bombers' fuselages were scorched black with soot.

Tokyo was devastated. The Japanese authorities had never prepared the city or its inhabitants to withstand a sustained aerial bombardment. Non-essential buildings had not been demolished to create firebreaks, nor had women and children been evacuated to the countryside, precautions that had been taken in most strategically important cities throughout Europe. The terrified citizens of Tokyo ran out into the open spaces and took refuge on islands and in water. But the fire had become so intense that it jumped the urban canals and the Sumida River, setting fire to everyone and everything in its path.

Right: The industrial area of Tokyo along the Sumida River after the attack. Only modern buildings of steel or concrete survived the concentrated low-level firebombing.

Below: American soldiers inspect the wreckage of Tokyo after the end of World War II. The people of Tokyo were quite unprepared for an air raid or the firestorm that followed.

At the height of the attack and for several hours afterwards, the glow from the inferno was visible 150 miles (240km) away. At its heart, those who were not burned to death were suffocated for lack of oxygen and had their lungs seared by poisonous fumes. The experience of Fusako Sasaki was typical of thousands: "As I ran, I kept my eyes on the sky. It was like a fireworks' display as the incendiaries exploded. People were aflame, rolling and writhing in agony, screaming piteously for help but beyond all mortal assistance."

In addition to the dead, 1.8 million people were injured or made homeless. Tokyo's hospitals were ill-equipped to deal with a disaster of this magnitude: they had no plasma, no painkillers, not even bandages. The all-clear was finally sounded at 0500 hours on the morning of March 10, but it took four days to put out the fire and 25 days to clear up all the bodies.

The Tokyo massacre had been designed to bring Japan to its knees and force it to surrender. But in the short term it had the opposite effect. The people of Japan – of all classes – became determined that they must fight on because they were now convinced that the Americans were bent on the destruction of the entire Japanese nation.

EMPIRE STATE BUILDING, NEW YORK, USA

JULY 28, 1945

In July 1945 with the war in Europe over, and the war in the east nearing its end, hundreds of aircraft began to arrive back in the United States from bases abroad. The United States produced no less than 11,000 B-25 medium bombers during World War II; when the war was over many of these ended their days as research platforms.

At 0855 hours on the morning of Saturday, July 28, Lieutenant-Colonel William Smith, an experienced flier with many hours under his belt, and a highly decorated combat veteran of World War II, took off from Bedford Field aerodrome, north of Boston. At 27 years old, he was one of the youngest lieutenant-colonels in the US Air Force, and had a reputation as something of a

Left: The Empire State Building, New York. The B-25 bomber plane piloted by Lieutenant-Colonel William Smith smashed into the 79th floor of the building, causing destruction and havoc to the city's landmark skyscraper.

Above: One of the 11,000 B-25 medium bombers produced for the United States forces during World War II. It was a B-25 that smashed into the Empire State Building on July 28, 1945.

Right: A US sailor holds up one of the bomber's propellers found in the wreckage.

daredevil, who was used to getting his own way. He had over 50 combat missions to his name on B-17 Flying Fortresses, but it was only the second time he had flown a B-25, which was lighter and faster. On this fateful morning Smith was flying without the benefit of a navigator.

Smith was scheduled to pick up some fellow army officers at Newark airport, New Jersey, where he had dropped them the previous evening. However, reports of bad weather at Newark meant that most of the aircraft landing at Newark that morning were requesting they be guided in using an instrument approach,

and this was causing a backlog in incoming traffic. In spite of this Smith decided to proceed and reached New York in about 50 minutes. He was initially refused landing permission because of the volume of traffic stacked up and waiting to land. Smith continued to cajole the heavily pressured air traffic controller at Newark airport to give him landing clearance, based purely on his ability to find his way down the glide path and onto the runway by sight.

A dense and impenetrable fog rolling in from the sea had reduced visibility in the New York area to a mere 220 yards (200m). As the B-25 crossed over New York, Smith apparently became disorientated in the foggy conditions, and office workers in New York's skyscrapers reported watching the hapless aircraft weaving through the towering concrete jungle. Without the benefit of either a cockpit voice recorder or a flight data recorder (both of which were not to be seen on the flight decks of aircraft until the mid 1960s) it is difficult to know how Lieutenant-Colonel Smith strayed so perilously from his flight path.

At about 1000 hours the aircraft smashed into the 79th floor of the Empire State Building in New York, a height of some 1000 feet (305m). Wreckage plowed through the building, smashing through solid walls and spreading burning fuel across the floor. One of the engines severed the cable in a lift shaft, sending the lift and its female attendant plummeting 1000 feet (305m) to the basement. Incredibly, she was saved by the giant springs that had been placed in the basement for just such an eventuality.

Burning wreckage from the aircraft crashed on to the street below. Miraculously only 13 people were killed, including the three occupants of the aircraft.

BOURNE END, ENGLAND
SEPTEMBER 30, 1945

Left: Sifting through the tangled mass of the Bourne End disaster, brought about by a degree of confusion over a signal system which led to a violent derailment.

Below: A badly damaged engine involved in the incident is lifted gently from the scene of the accident.

T his derailment in the first few weeks after the end of World War II highlighted the potential dangers associated with signal warning systems and the possible confusion that even a railroad's most experienced drivers might face in interpreting them. On this occasion, an overnight express of the London, Midland and Scottish Railway bound for London's Euston station from Perth in Scotland was derailed when its driver failed to negotiate a section of track at the approved speed.

The incident at Bourne End began as the express was crossing over a section of track that linked the fast and slow lines. The switch from the express's usual fast track had been made necessary by weekend repair work on the up line to London. The driver should have been aware of this potential hazard as the details had been published in a fortnightly briefing document and circulated to all the railroad's staff. There was a 20mph (32km/h) speed restriction in force along the section of

Above: This aerial view of the disaster site gives a good indication of the damage inflicted on both the rolling stock and track at Bourne End.

track in the vicinity of the work, although a train could have crossed the point at double this speed and not suffered any undue consequences.

However, the Perth–London up express was traveling at 50mph (80km/h), too high a speed to prevent the fatal derailment. The "Royal Scot" locomotive left the track with great momentum and crashed down to the bottom of the high embankment it was negotiating. It was followed by seven carriages which landed on top of the locomotive with considerable force. Five more of the express's carriages were also derailed and only three stayed on the line. The violence of the initial derailment and the subsequent drop down the embankment left 43 people dead and more than 120 injured.

Although the driver had clearly been traveling far too fast to prevent the accident, investigators suggested that his failure to reduce speed may have been the result of him failing to interpret confusing signals correctly. Recently installed signals some way from the point where the express had to switch to the slow track should have given the driver ample warning of the danger ahead.

These were placed some 2600 yards (2400m) from the point at which the track switch was to be made. However, in other sections of track, a signal similar to that at Bourne End would have preceded a second signal, one giving a final warning of the hazard.

It was possible to argue that the driver had assumed that this second type of signalling system was in operation at Bourne End and that, consequently, he believed he had far more time to make the necessary reduction in speed than he actually had. In this case, driver error had been brought about by possible inconsistencies in a network's signalling policies. Investigators recommended a thorough overhaul of warning practices on the rail network.

NAPIERVILLE, ILLINOIS, US

APRIL 25, 1946

The causes of some disasters are not always clear cut, as this incident shows. The driver of the train in question was indicted for manslaughter but the case was dropped because of insufficient evidence, yet there was no indication of any mechanical failure to explain the crash, even after signalling systems and the train itself were examined.

Below: Rescue crews use ladders to gain entry into one of the carriages damaged at Napierville, while colleagues remove one of the dead.

This crash, which left 45 passengers dead and 36 injured, began when the conductor of the westbound "Advance Flyer" spotted an object thrown out from below his train. His driver then halted the train at the next station, Napierville, where the train's conductor dismounted and walked some 300 yards (276m) back down the track to watch for on-coming traffic. He immediately spotted the "Exposition Flyer" approaching, which applied its brakes, but hit the stationary "Advance Flyer" at approximately 50mph (80km/h).

The impact was such that the second locomotive ploughed through the last carriage of the stopped train. The next carriage along suffered mild damage, but the following dining car was crushed into a U-shape, partly because of its lightweight construction.

Signalling in the area of the crash was examined for faults, but none was found, and the driver of the "Exposition Flyer" did have sufficient warning to apply his emergency brakes, but there was no indication that he had done so. While there was no clear-cut evidence of human error, the lack of credible alternative explanations continues to make it the most likely cause of the Napierville accident.

WINECOFF HOTEL, ATLANTA, USA

DECEMBER 7, 1946

A total of 137 people were killed on the night of December 7, 1946, when fire engulfed the 15-story Winecoff Hotel in the heart of the business district of Atlanta, Georgia, USA. As many as 116 were pronounced dead at the scene; another 21 people died later in hospital from agonizing burns or the effects of smoke inhalation.

When any tall building catches fire, people on the upper floors are faced with a terrible choice. They can either jump immediately, risking almost certain death, or wait inside the building in the hope that help will reach them before they are engulfed in flames or overcome by smoke.

In the Winecoff tragedy, the death toll was highest among those who stayed in their rooms – almost 100 perished in this way, including the owner of the hotel, 70-year-old W.F. Winecoff, who had built the hotel in 1913 and had latterly been a resident.

But many of those who left by the windows also died, including 25 people who jumped from the upper storys and were killed by the impact on landing. Some others tried to climb down ropes made out of bedsheets which had been hastily tied together – too hastily, as it turned out. The sheets were not knotted securely enough to take the weight and these people too fell to their deaths.

Most of those who survived were rescued by firemen – some climbed down ladders which had been cranked up to the high windows, while others leaped from as high as the tenth floor into outstretched jump-nets made of rope.

Left: Firemen direct streams of water through high-powered hoses into the upper floors of the Winecoff Hotel, situated in Atlanta's famous Peachtree Street. Many of the occupants of the 350 rooms in the hotel were asleep when fire swept through the building in the early hours of a Saturday morning. Of those who died in the fire, a number were killed when they jumped to their deaths from the upper storys of the hotel.

GRAND CAMP, TEXAS CITY

APRIL 16, 1947

A port can be a dangerous place. When ships carrying combustible or explosive cargoes are docked alongside each other, a single outbreak of fire can lead to a chain reaction. This was the case in the horrific disaster at Texas City in April 1947. When a fire broke out on the French freighter *Grand Camp* while it was in dock, it resulted in an explosion that wrecked numerous other ships, destroyed a chemical plant, and left an estimated 90 percent of Texas City, a key port on the Gulf of Mexico, in ruins. Some 800 fatalities were recorded by the Red Cross, but the final death toll was probably much higher. The area was declared a disaster zone, so great was the catastrophe.

Texas City was a major terminal for oil tankers, but the *Grand Camp* was carrying a potentially more lethal cargo – highly combustible ammonium nitrate, a white crystalline solid used in the manufacture of fertilizers and explosives.

Fire broke out on the *Grand Camp* in the early morning of April 16 and spread so quickly that the crew had no time to alert the harbor authorities. The fire took a firm hold on the vessel and ignited the ammonium nitrate in its hold. The chemical vaporized and produced an enormous explosion that obliterated the *Grand Camp*, flinging burning debris high into the air and across the harbor. The burning debris landed on the various buildings in the harbor and on many of

Below: Huge plumes of black, toxic smoke still linger over the port of Texas City the day after the Grand Camp *with its cargo of ammonium nitrate exploded.*

Right: Two women kneel to pray in the remains of the shattered Catholic church in Texas City. About 90 percent of the city was destroyed by the series of huge explosions set off by the Grand Camp.

Below: An emergency first aid station was set up to treat victims of the explosion.

the 50 or so oil tankers that were loading oil at the time. These tankers also began to burn out of control, spreading flames even farther afield. Thick plumes of acrid smoke soon blanketed the city in toxic fumes.

The fire continued to spread, quickly reaching a chemical plant. The city's overstretched fire department and rescue services were simply swamped by the magnitude of the disaster that confronted them. They tried hard to prevent the fires from spreading but with little success – the blaze continued out of control for several days. On the day after the initial explosion and fire, a second cargo vessel, the *High Flyer*, also blew up, adding to the carnage.

Much of Texas City was left in ruins. The initial blast flattened many of its wooden and brick buildings and the subsequent fires destroyed many more. It was even reported that a two-seater light aircraft flying over the harbor at the moment of detonation was a victim of the catastrophe. So powerful was the first explosion that there were reports of windows being smashed more than 10 miles (16km) from the point of detonation.

The devastation was horrific. A combination of explosive and combustible materials had torn the heart out of the city. It took many years – and millions of dollars – to restore the city to its position as one of the major ports on the Gulf of Mexico.

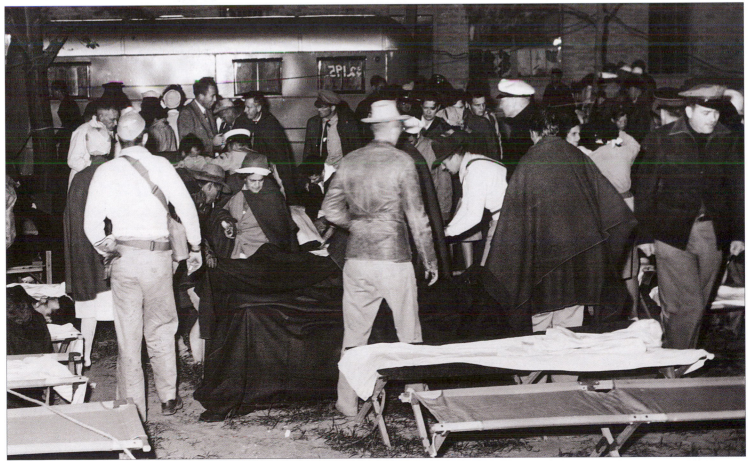

SOUTH CROYDON, ENGLAND

OCTOBER 24, 1947

This remains one of the most catastrophic accidents brought about by the negligence of a rail employee to take place on the British rail network. The incident happened on a foggy morning at the height of the rush hour as workers from southeast England were heading for their offices in London. Thirty-two people were killed and 183 injured in a collision near South Croydon that involved one train traveling between Tattenham Corner and London Bridge station and another also heading for London from Haywards Heath.

The accident was brought about by the Tattenham Corner train receiving an incorrect signal which indicated that the way ahead was clear. As it reached a

speed of over 40mph (64km/h), it hit the back of the Haywards Heath train, which had reached a speed of 20mph (32km/h) shortly after it had been given the all-clear to continue its journey.

The crash was brought about by a misunderstanding between two signalmen operating somewhat antiquated equipment. Semaphore signals were still being used on this section of busy track, rather than the newer colored lighting system. A signal for a train to proceed on its way given by one signalman could only be given if the signalman in the next box up the line accepted the signal.

Once the second worker had accepted the signal, a blocking lever could not be replaced until the train in question had activated a device on the track a short

Above: Over 30 people lost their lives in this collision at South Croydon brought about by a lack of understanding between two signalmen.

Right: A trackside view of some of the coaches that were destroyed in the two-train smash. Fog still blankets the crash site.

Below: A doctor and official emerge from beneath the remains of one of the carriages destroyed by the impact between the two trains.

distance from the signal. This rigid system of signals and blocks could be overruled by a signalman acting on his own initiative if one of the system's components failed for some reason, but the railroad authorities had laid down very strict procedures for tackling such unusual events.

On the day in question, a signalman manning the box at Purley Oaks forgot about the train from Haywards Heath, which had been waiting in the station for more than five minutes. His view was obscured by the thick fog that was blanketing the area, and which may have contributed to his misunderstanding of the situation. When he received a request from the next signalman down the line asking him if it was permissible for the Tattenham–London Bridge train to proceed, he thought that the Haywards Heath train had left the station, but mechanical failure in the signal system had prevented him from being informed of the fact. Consequently, he overrode the system without following the formal guidelines for dealing with such an event and gave his permission for the second train to proceed on its way.

Because of this serious breach of regulations, the second signalman was able to give the signal for the Tattenham–London Bridge train to continue its journey. A few moments later, the stationary Haywards Heath–London service was hit in the rear.

Perhaps due to luck or the hazardous conditions which forced trains to reduce their speeds, the incident was not as severe as might have been the case. Less than a third of those injured in the collision needed hospital care, and of these 58 only 41 had to be kept in for treatment. Nevertheless, 32 passengers had lost their lives at South Croydon because an employee failed to follow safety regulations.

WINSFORD, ENGLAND

APRIL 17, 1948

Left: Rail engineers and investigators attend the site of the Winsford crash, which involved a pair of trains heading for London's Euston station from Glasgow.

This collision just north of Winsford Junction station, part of the British state-owned network's London Midland Region, was brought about by the actions of an irresponsible passenger and an error by a British Railways' official. Their unfortunate actions caused an accident that left 24 dead and nearly 20 injured.

On the evening of April 17, an express from Glasgow in central Scotland bound for London's Euston station was brought to an unscheduled halt by a British soldier who pulled an emergency communications cord in one of his carriage's toilets. Almost unbelievably, he did this not because he was in any personal difficulty, but because he wanted to reach home as quickly as possible. Rather than travel a little farther down to Crewe station, change to another train there, and then head back to his home station, he halted the London-bound express with the intention of jumping off.

Meanwhile, at the Winsford signal box the British Rail employee on duty there mistakenly believed that he had seen the Glasgow–London express go by some

time earlier and, on this false assumption, allowed a following postal train, also from Glasgow, to continue on its way south to the capital. His "train out of section" message to the postal train sealed the fate of the stationary Glasgow–London express ahead and its many passengers.

Back at the stationary express, which had been held up for a little less than 20 minutes by the soldier's pulling of the communication cord, a guard had gathered up his red warning lantern, left the train at the rear, and set off back down the track to warn any following traffic on the same line that his express was waiting to proceed after its unscheduled halt.

The guard had walked about 400 yards (370m) from the Glasgow–London express when he spotted the advancing postal train in the near distance. By waving his red warning light as laid out in safety manuals, he was able to slow the on-coming postal train, but the reduction in its speed was insufficient over such a short distance to bring the train to an emergency stop and prevent a fatal rear-end collision with the stopped express. Moments later, the crash occurred. It was a

Left: As a crowd of spectators looks on, heavy lifting equipment is brought up to remove the remains of the two trains smashed in the incident at Winsford.

Right: The engine of the post train from Glasgow lies embedded in the wreckage of one of the halted express's rear carriages.

little after midnight. The last carriage on the Glasgow–London express was totally destroyed in the violent smash, and the adjacent carriage was also badly damaged. Both contained several passengers.

In the subsequent inquiry, it was stated that the Glasgow–London express's guard might have prevented the accident if he had been quicker in leaving his train after the cord had been pulled by the soldier and then moved farther down the track, thereby giving the second train a greater distance to come to a complete halt. However, it was clearly stated that the wanton foolishness of the off-duty soldier and the mistake of the Winsford Junction signalman were the prime causes behind the fatal collision.

Most people desired a return to normality as quickly as possible after the end of World War II, but the conflict's impact was destined to be felt for many years after 1945, not least in the continued use of outdated and worn-out technology that many financially constrained countries could not easily replace. Equally, as standards of living began to rise for those in the more affluent nations, so too did the desire to travel for recreational reasons. Many ex-servicemen, who had served overseas, had retained an interest in foreign travel. However, ships were too slow for those in a hurry or with limited holiday entitlements, and so aircraft began to corner the steadily growing market in recreational travel.

Air travel remained a novelty during this period, one available to few, as mass air transport only really developed in the 1960s. However, the early pioneers of holiday air travel had to face several potential dangers, not least outdated machines, poor safety regulations, and often badly organized air space. Mixed together these could be a recipe for disaster and on occasion were. The loss of 65 passengers and crew out of a total of 73 people on board an American Airlines' Lockheed Electra in New York's East River in February 1959 indicated that air crashes could be devastating. Nevertheless, aviation catastrophes remained unusual, and newspapers were more likely to be filled with stories concerning fires, sinkings, and train crashes.

Background: *The scene at Harrow and Wealdstone station, England, on October 8, 1952, following a crash involving three trains. More than 100 people died and 150 were injured.*

1950–1959

ROCKVILLE CENTER, NEW YORK STATE, USA

FEBRUARY 17, 1950

This incident on Long Island resulted in the deaths of 31 commuters. Rockville Center was undergoing major engineering work at the time of the collision. Railroad workers were separating the levels of the up and down tracks, but the only way to accommodate existing traffic during the construction period was to run trains on a single section of track for a short distance. Automatic signals were placed at each end of the single-track section and decisions on train priority were the responsibility of Rockville Center station.

The collision occurred shortly after 2230 hours when a commuter train traveling to New York from Babylon was given right of way on the temporary section of track by those on duty at Rockville Center. Signals were switched to halt any train moving in the opposite direction from entering the stretch of single track. The journey progressed smoothly enough until the Babylon–New York train was on the point of leaving the single-track and returning to its own line.

At this moment, however, it was struck by a train also packed with passengers, many returning home after a night out, which had gone through the protecting signals. Both drivers survived the 50mph (80km/h) collision, which saw the two trains strike each other at a slight angle, because their cabs were positioned away from the point of impact, but commuters on both trains died when the lefthand sides of the carriages they occupied were crushed.

An enquiry revealed that the driver of the train responsible for the incident had seen the stop signal and heard a warning buzzer sound in his cab, but had suffered a momentary loss of consciousness due to a medical condition. His version of events was accepted by the investigators.

Below: The scene at Rockville Center, New York State, shortly after the fatal impact between two trains. One of the injured is taken out by stretcher while rescue workers try to free the trapped.

SIGGINSTON, WALES

MARCH 12, 1950

Avro's Tudor aircraft was one of the earliest airliners. It was developed and first put into service in the mid 1940s. Its history was marred by several crashes, one of which killed its designer, Roy Chadwick. On March 12, 1950, another Tudor V aircraft crashed with a party of Welsh rugby football supporters on board.

The crash of the Tudor V was due to errors both by the ground crew, and by the pilot failing to make adequate preflight checks regarding the correct distribution of weight in the aircraft.

The aircraft was chartered by a group of Welsh rugby fans to ferry them to Dublin in Ireland, where they watched Wales beat Ireland in the international match at Belfast. On the day following the Welsh victory, the jubilant fans boarded the aircraft at Dublin for the homeward flight back to Llandow airport, serving Cardiff.

For the return flight, six more passengers had been added to the manifest. Crucially, although the seating arrangements had been altered – thus affecting the balance of the aircraft – the luggage in the hold was neither sufficiently heavy, nor properly distributed, to correct the center of gravity.

At about 1445 hours, as the aircraft was making its final approach to Llandow, observers on the ground realized that it was in trouble. When the Tudor was over the runway center line at a height of about 120 feet (37m) the pilot tried to correct his angle of descent (which investigators adjudged to have been too steep) by applying full power. However, because the poor distribution of weight had resulted in a rearward shift of the center of gravity, this action made the aircraft difficult to control.

The Tudor climbed rapidly to a height of about 300 feet (100m), stalled, and crashed into a field, coming to rest only 20 yards (18m) from a farmhouse in the little village of Sigginston. Although there was no explosion, the impact killed all 83 of those on board.

Left: The scene just outside the village of Sigginston, South Wales, the day after the Avro Tudor V crashed while making its final approach for landing at Llandow airport.

RICHMOND HILL, NEW YORK STATE, USA

NOVEMBER 22, 1950

Collisions at night are not an uncommon phenomenon on the world's railroads, as this incident which resulted in 79 deaths and 352 injured confirms. The crash in question involved a pair of 12-car electric trains. One was traveling from Penn station in New York to Hampstead station when a signal in the vicinity of the town of Jamaica indicated that it should reduce speed. However, when given the go-ahead to proceed as normal, the train's brake would not release fully.

The train shuddered to a halt because of the braking fault and, as the regulations demanded, the conductor disembarked with a warning lantern to keep watch at the rear of the train. Because the signal system in use on this section of track did not permit any halted train

getting under way to travel at more than 15mph (24km/h) when given the "stop and proceed" signal until the next clear signal was seen, the conductor did not move any great distance down the track from his train. He expected that any following train would be obeying the same 15mph (24km/h) rule and would not, therefore, require as great a stopping distance as if it was moving at a much higher speed. It was to be an unfortunate but understandable miscalculation.

The conductor soon returned to the stopped train when he heard the driver attempting to move off. However, the train remained stationary because of the problem with its braking system. Moments later, the conductor had little time to do more than wave his red lantern in urgent warning as a second train loomed out of the darkness, advancing at high speed. Unable to

Below: A pair of cranes begin to gently lift one of the coaches crushed by the collision at Richmond Hill. At this stage, a number of victims were still trapped in the wreckage.

brake in the available distance. It ploughed into the stationary train, driving it forward by some 25 yards (23m) and causing multiple casualties among the passengers on both trains.

It was later revealed that the second train had also received a 15mph (24km/h) "stop and proceed" signal. Its driver had followed the standard procedure correctly but then gradually picked up speed as he advanced. It was discovered that the second train was moving at 35mph (65km/h) when its driver attempted to engage his locomotive's emergency brake in a futile attempt to prevent the accident from occurring.

The driver in this following train was killed in the accident and investigators were never able to fully explain the reasons behind the increase in speed of the following locomotive that sealed the fate of the stationary train in front.

It was suggested that the second driver had misinterpreted the clear signal that should have allowed the stationary train in front of him to proceed as normal had it been able. He may have thought that it was directed at him, not the train in front, and, therefore, allowed him to increase his speed to normal, thereby setting in motion the chain of events that led to the collision between the two.

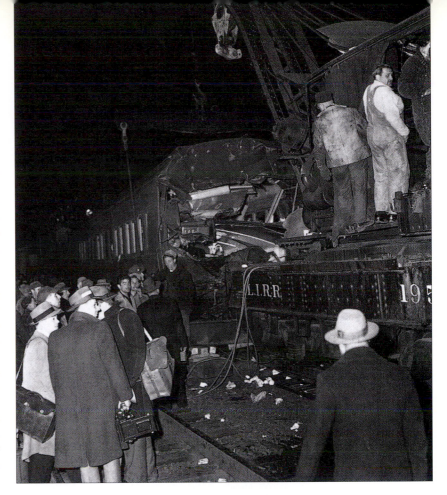

Above: *Rescue workers continue the grim task of removing the living and dead from the remains of one of the coaches smashed in the collision.*

Left: *Doctors and emergency crews begin the emotionally draining task of recovering bodies from the tangled wreckage of one of the two trains.*

WOODBRIDGE, NEW JERSEY, USA

FEBRUARY 6, 1951

The official inquiry by the New Jersey Board of Public Utility Commissioners that followed this crash that left 84 people dead and many injured blamed the incident on a driver who had exceeded the speed restrictions on a potentially dangerous section of track a little to the west of Woodbridge station in New Jersey. Early in the afternoon of February 6, a speed restriction was introduced on the section of track in question, where a length of the New Jersey turnpike was being overhauled and the track had been curved to allow the contractors to carry out their renovations. The driver of "The Broker," a steam-powered commuter train packed with passengers because of a strike on the Jersey Central line, reached the 25mph (40km/h) restriction zone at a speed of 60mph (96km/h), thereby sealing the fate of the passengers.

The locomotive was able to pass the first section of temporarily curved track without any difficulty, but as it entered the second part of the curving section, its weight was thrown on to the wrong rail for negotiating

Below: An aerial view of the crash scene gives a good indication of the scale of the disaster. Two carriages lie on the edge of the bridge.

the curve. The locomotive was tipped over on to its righthand side, coming to rest on the top of the embankment along which it was traveling.

Seven of the all-steel carriages it was hauling were also derailed. Of these, four were scattered across the tracks, while one, the last of the seven, had been ripped open along the entire length of one side. Another coach came to rest jammed into a bridge, parts of which collapsed under the violence of the enormous impact.

The board of inquiry established that the driver had been aware of the speed restriction before he started the journey as he admitted to having read a special note to that effect posted by the railroad in his New Jersey depot. However, there was some question as to the precise section of track along the route to which the warning applied.

The route he was following, between Jersey City and Bay Head, made use of two different railroads with separate safety procedures. One section, that up to Perth Amboy, was run on tracks owned by the Pennsylvania Railroad, while the second, longer, stretch belonged to the New York and Long Branch. On the shorter stretch, speed restriction signals were not marked, but they were on the second.

Giving testimony at the subsequent board of inquiry, the driver claimed that he had been keeping watch for the relevant danger signals. As it was his first journey since the 25mph (40km/h) speed restriction at Woodbridge had come into force, he may have been lulled into a dangerously false sense of security on the early part of his journey and missed the subsequent danger signs.

Although the driver was held directly to blame for the accident by the board of inquiry, the lack of consistent safety regulations on the two sections of track may have contributed to the circumstances which led to the derailment of his train.

Below: Several floodlights illuminate the crash site as rescuers sift through the coaches in search of survivors.

Right: Two cranes attempt to right an overturned locomotive 48 hours after the disaster at Woodbridge.

HARROW AND WEALDSTONE, ENGLAND

OCTOBER 8, 1952

Below: Rescue workers aided by heavy equipment struggle to search through the tangle of wreckage created by the three-train smash at Harrow and Wealdstone station in early October 1952.

This collision remains one of the worst incidents ever to have taken place on the British railroad network. The crash between a packed morning commuter train on its way from Tring in Hertfordshire to London's Euston station, an overnight sleeper train from Perth in Scotland to London, and a twin-locomotive express running from Liverpool to Euston left 112 people dead and over 150 injured, many seriously.

The commuter train had been waiting at Harrow and Wealdstone station in north London for about a minute when it was struck in the rear with massive force by the sleeper train from Scotland. The impact sent wreckage flying on to the line that carried the Liverpool–Euston express, which careered into the obstruction just a few seconds later. The express's two locomotives were traveling at such high speed that they mounted one of the station's platforms, blocking the local electric line. Sixteen carriages suffered massive damage; many of them were reduced to an unrecognizable tangle.

The driver of the Perth–London sleeper was subsequently blamed for the disaster as it was stated that he failed to heed signal instructions. His 11-coach train, pulled by the "City of Glasgow," a "Princess Coronation" class locomotive, was more than 30 minutes behind schedule, partly due to fog, when the incident

Above: The shattered body of a locomotive involved in the disaster. Despite serious damage, one of them, the "City of Glasgow," was repaired and put back into service.

notice or act on the warnings and continued at speed toward Harrow and Wealdstone.

Shortly before the arrival of the Perth train, the Tring–London passenger train, consisting of eight carriages, had been switched from the slow line to the fast line just north of the station by a signalman as usual, prior to reaching the station's platform. Consequently, the fast line's signals were in the danger or caution position but were ignored. The driver of the "City of Glasgow" braked at the last moment but the collision could not be avoided. After the first collision had taken place and wreckage blocked the path of the Liverpool–Euston train, the second devastating impact was an absolute certainty.

The local emergency services were swamped by the scale of the accident and their systems for dealing with a crash of this magnitude were shown to be inadequate, despite their best efforts. However, US military personnel drafted in from nearby bases implemented the triage system of assessing casualties on the basis of the extent of their injuries and did much to prevent an even heavier loss of life. The system was subsequently adopted by the British emergency services. The tragedy at Harrow and Wealdstone also provided further impetus for the introduction of the automatic warning system, which sounds a siren in a driver's cab when his train approaches a signal showing caution.

occurred, but conditions at Harrow and Wealdstone station were fairly clear. A signalman spotted the "City of Glasgow" approaching when it was some 600 yards (554m) from his box and had time to lay down stop detonators on the track by pulling the relevant lever. For reasons never fully explained, the driver failed to

Right: Once the casualties had been evacuated, it was essential that the station was opened as quickly as possible. Here, rail workers and firemen prepare to remove a wrecked carriage.

CONNEAUT, OHIO, USA

MARCH 27, 1953

Below: One of the locomotives derailed in the Conneaut collision lies on its side. One of the pipes that played a key part in the smash can be seen in the bottom left section of the photograph.

This incident was brought about by the failure of railroad inspectors to check the fastenings that held a cargo of large pipes in place. These had, in fact, been loaded incorrectly at their point of origin – a steel works in Aliquippa, Pennsylvania. Some of the pipes were placed above the sides of the wagons on to which they were loaded. These two errors led to 21 deaths in the subsequent collision which involved three trains.

The pipes, each over a foot (30cm) in diameter and more than 30 feet (9.2m) long, had been placed on wagons pulled by a freight train heading eastward along a main line consisting of four tracks. As the freight train was making its journey, movements were sufficiently great to further loosen the fastenings and one of the large pipes worked its way free, becoming lodged between a wagon and an adjacent track.

The impact of the pipe as it was dragged along led to the second track being pushed out of position by approximately 18 inches (45cm). A westbound train traveling in excess of 70mph (112km/h) was derailed by the distorted track as it was passing the freight train into which it collided, scattering wreckage across the rail lines. This wreckage on the track sealed the fate of the eastbound "Southwestern Limited," also traveling at 70mph (112km/h). With the exception of its last three carriages, the "Southwestern Limited" was derailed. Seven of the passenger coaches from the two passenger trains were totally destroyed in the incident.

WEST BENGAL, INDIA

MAY 2, 1953

Above: A BOAC Comet 1 similar to the aircraft that crashed soon after take-off from Calcutta on May 2, 1953. The airliner was powered by four De Havilland Ghost engines, and carried a flight crew of four, plus two cabin crew, with accommodation for about 40 passengers.

This accident was the third in a series of seven mysterious accidents involving a De Havilland Comet aircraft in the space of 18 months. In April of 1954, all Comets in service were grounded, due to serious questions about the structural integrity of the aircraft.

On the first anniversary of the inauguration of commercial jetliner flights, a Comet 1 in the service of British Overseas Airways Corporation (BOAC) took off from Singapore on a westbound scheduled service to London Heathrow. Captaining the aircraft was one of the more experienced jet pilots in the service of BOAC. Only two months previously a brand-new Comet had been written off at Karachi by a pilot with little or no experience on the type.

Captain Haddon was rather more capable, and handled the take-off (notoriously difficult to judge accurately on the Comet) with precision. The Comet completed the first leg to Calcutta without incident. At Calcutta passengers for Delhi and London – the second and final destinations of the service – were taken on

board. It was just before 1630 hours when the Comet took off from Dum Dum airport at Calcutta. The flight crew began to climb to their cruising height for the transit to Delhi, 800 miles (1287km) to the northwest. Outside, a violent tropical thunderstorm was raging.

Six minutes after take-off, as the Comet was ascending through 10,000 feet (3000m), the aircraft broke up in the air. The wingless Comet plummeted to the ground in flames and exploded in a forest near the village of Jugalgari, some 30 miles (48km) northwest of Calcutta. All 37 passengers and six crew members were killed.

An extensive investigation was immediately launched, both by the Indian government and at the Royal Aircraft Establishment (RAE) at Farnborough, England. Powerful winds had carried wreckage over a large area, and it took many months before an accurate analysis of the incident could be made.

The Indian inquiry concluded that the strength of the turbulence had placed loads on the airframe of the Comet in excess of those which it had been designed to withstand, and of a strength that would have caused

Above: The wreckage of the crashed Comet strewn over the ground near the village of Jugalgari, north-west of Calcutta.

the disintegration of any commercial transport aircraft. There were, however, indications that metal fatigue could have played a role in the accident. However, it proved impossible to substantiate these claims, because the Indian authorities mistakenly disposed of much of the wreckage before the RAE could examine it.

The British inquiry focused on the likelihood that the airframe had been fatally overstressed as Captain Haddon tried to pull out of the dive that the Comet entered when a spar in the port elevator snapped. The British Air Registration Board, who at that time

granted certification to any new British aircraft types, called for more tests to establish susceptibility of the Comet airframe to metal fatigue.

During the following six months a Comet aircraft was subjected to extensive fatigue loading tests that failed to show any evidence that the aircraft was liable to failures.

However, the number of incidents involving the Comet during those crucial early years of its operational life did serious damage to potential export sales and led to the cancellation of many orders.

WAIOURU, NEW ZEALAND

DECEMBER 24, 1953

Below: The site of the crash which carried away the Wellington to Auckland express. Floodwaters caused by a volcanic eruption swept away the bridge seen here and the train plunged into the turbulent waters.

This accident must rank as one of the most unusual to have ever occurred in New Zealand as it was caused by the eruption of Mount Ruapehu, a volcano on the North Island. The Wellington to Auckland express was packed with passengers heading home to begin the Christmas festivities. The train passed through Waiouru on time and all seemed well.

Unbeknown to the express's passengers and crew, a bridge ahead of them crossing the Whangaehu River was about to be washed away. The eruption of Mount Ruapehu had released the waters of the lake in the volcano's crater, which proceeded to flow into the river valley through caves in a glacier. The massive volume of water traveling at high speed tore away the river banks, picking up debris as it headed for the bridge.

The waters swamped the bridge as the train approached, sweeping away its fourth pier and damaging the fifth. A motorist tried to raise the alarm before the train reached the bridge and the train's driver applied the brakes at the last moment but to no avail. The train roared on to the bridge and then plunged into the foaming waters along with six carriages.

Of the 285 people on the train, 134 survived the incident and of the 151 who died, 20 bodies were never recovered. Eight others were buried without being identified. So much force had been generated by the floodwaters that the remnants of one carriage were found more than five miles (8km) from the bridge.

GRAND CANYON, ARIZONA, USA

JUNE 30, 1956

The huge expansion in the popularity of air travel during the 1950s placed a great burden on the limited capacity of US air traffic control. Despite the introduction of professional air traffic controllers in 1929, the system had not kept pace with the increasing distances and greater volumes of human traffic on internal US routes. The United Airlines–TWA collision in June 1956 proved not only the most costly civil air accident in human terms of that decade, but also highlighted the woeful inadequacies of this system.

The Super Constellation (the TWA aircraft involved in the crash) was introduced into service in 1951. The Douglas DC-7 (the United Airlines aircraft) was designed following a request by American Airlines for a commercial competitor on US trunk routes to the TWA Super Constellations.

In the late morning of Saturday, June 30, the TWA Super Constellation took off in fine conditions from Los Angeles International airport, California, on a scheduled service to Kansas, and eventually Washington, DC. It was followed a few minutes later by

Left: The Grand Canyon in Arizona is a spectacular feature. The passengers of the TWA Super Constellation and the United Airlines DC-7 were probably craning to catch a glimpse of the breathtaking view when their aircraft collided, killing all on board.

the two aircraft were set to cross over the Grand Canyon. Once out of the jurisdiction of the Los Angeles air traffic control, the two aircraft were in uncontrolled airspace, and they would be operating under visual flight rules. This meant that it became the sole responsibility of the flight crew to avoid other aircraft in the area. However, the air traffic controller responsible for both flights was at fault for failing to inform either crew of other aircraft movements in the area.

With nothing to warn them of the impending collision, the passengers of the TWA aircraft were probably gazing down through the cloud hoping to catch a glimpse of the spectacular scenery of the Grand Canyon when the collision occurred. The DC-7 was descending – ironically, the DC-7's captain was probably trying to provide his passengers with a better view. The rear fuselage of the Super Constellation was torn off in the impact, which also destroyed the outer section of the port wing on the DC–7.

With no means of controlling the aircraft, the TWA Constellation plummeted almost vertically upside down into the canyon. The DC-7 impacted about a mile away on the steep slopes of the canyon wall. All 58 people on board the DC-7, and all 70 of those on the TWA Constellation were killed.

Above: An aerial view showing (upper arrow) where the wreck of the DC-7 was found and (bottom arrow) where the Constellation was found.

Right: Pieces of the Constellation strewn all over the side of the mighty Grand Canyon.

the United Airlines DC-7, bound for Chicago, Illinois, where it was due to make a stopover before completing the final leg of its journey to Newark, New Jersey.

The faster DC-7 climbed to a cruising altitude that was slightly higher than that of the Super Constellation and both aircraft headed out across the Mojave Desert. Encountering cloud above the desert the pilot of the TWA Constellation requested permission from the Los Angeles air traffic control center to climb to 21,000 feet (6400m), but as the United Airlines DC-7 was flying in the same direction at this height the Constellation's request was denied.

Fatally, however, the Constellation's pilot was cleared to fly above the cloud layer – which also meant flying at approximately 21,000 feet (6400m). As they were flying on slightly different headings, the paths of

ANDREA DORIA, ATLANTIC

JULY 25, 1956

The *Andrea Doria* was a luxurious, well-appointed liner with modern furnishings that was an immediate hit with its passengers when it began its maiden voyage on January 14, 1953. Its owners, Italia-Società per Azioni di Navigazione, had high hopes for the ship. Yet its life was to end three and a half years later in a shocking collision in thick fog.

The *Andrea Doria* sailed between Genoa and New York, with scheduled stops at Cannes, Naples, and Gibraltar. By the middle of 1956 the *Andrea Doria* had made 50 transatlantic crossings. However, its 51st voyage was to end in catastrophe.

Captain Calamai, the master who had sailed with the *Andrea Doria* on its first voyage, was in command on this trip. As the liner headed out into the Atlantic it had a crew of 572 on board plus 1134 passengers enjoying the ship's facilities, which included three outdoor swimming pools and many public rooms decorated in modern Italian style. The crossing of the Atlantic went without a hitch until the ship was one day out from New York. It was July 25.

The *Andrea Doria* approached Nantucket lightship at full speed but thick fog had reduced visibility to less than half a mile by mid-afternoon. Calamai ordered the liner's speed to be reduced to 21 knots. Shortly after 2230 hours the *Andrea Doria*'s radar operator spotted an object ahead and the ship's officers calculated that the object would pass the liner on its starboard side. Calamai ordered that the ship's fog horn should be sounded every 100 seconds.

Above: The Andrea Doria *was a luxury liner plying between Italy and New York.*

Calamai ordered his officers to listen out for sounds from the other ship, but they suddenly spotted lights heading directly for them out of the mist. The *Andrea Doria* sent out two warning blasts from its fog horn and attempted to make a sharp turn. However, weighing in at nearly 30,000 gross tons, the *Andrea Doria* was slow to respond to the helm. The second vessel, the Swedish American Line's *Stockholm*, smashed into the *Andrea Doria*'s starboard side.

The damage inflicted was colossal. The *Stockholm* was traveling at 18 knots and had a specially strengthened bow designed to cut through ice. This bow sliced into the *Andrea Doria*'s hull to a depth of 30 feet (9m), the gap running from the liner's upper deck to well below the waterline.

The *Stockholm* was able to reverse engines and pull away from the *Andrea Doria*, which took on an immediate and fatal list. Many of the liner's lifeboats could not be launched because the ship was listing so badly, so the captain delayed broadcasting the order to abandon ship. Instead, he sent out a distress call, which was answered by a number of nearby vessels. These vessels were quickly on the scene and rescued over 1600 of the *Andrea Doria*'s passengers and crew.

The *Stockholm*, although suffering considerable damage to its bow, was also able to aid the rescue effort. When the survivors were counted, it was found that only 47 people had been lost, and most of these had been killed by the initial impact. The *Andrea Doria* sank on July 26, the day after the collision.

Above: *The tangled mass that was the bow of the damaged* Stockholm, *which limped into New York carrying 500 survivors from the* Andrea Doria.

Left: *After the impact with the* Stockholm, *the* Andrea Doria *keeled over to starboard, and the following day sank in 225 feet (68m) of water.*

KUURILA, FINLAND

Above: Finnish emergency crews and investigators recover the dead from the Kuurila disaster site in the south of the country.

Many of the world's rail networks operate in the face of the worst types of weather imaginable and, as this incident indicates, no matter how efficient the system, those in charge are sometimes caught out by atrocious conditions. On this occasion, over 25 people died and 45 were injured in a collision in part brought about by a severe blizzard.

Two expresses traveling at high speed in opposite directions were making their way through the blizzard to their destinations. Some 80 miles (128km) north of the capital, Helsinki, they both entered a section of single-line track at the same time.

A little later, they collided violently, and the engine and leading carriages of each train were badly damaged due to the resulting derailment. Several had to be scrapped as they had been reduced to little more than shattered shells. The incident was the worst disaster the Finnish rail network had experienced since the end of World War II.

To make matters much worse for both the survivors and the rescue services, the remoteness of the crash site, amid a dense pine forest covered in deep winter snow, delayed aid reaching the stricken trains and their passengers. Rescuers resorted to the use of skis and horse-drawn sleighs to reach the trains and remove some of the casualties.

PAMIR, ATLANTIC

SEPTEMBER 21, 1957

The loss of the *Pamir* led to an outpouring of grief among the public of West Germany when news of its sinking was reported by the media. The 3103-ton *Pamir* was being used as a training vessel and was in the Atlantic, bound for Hamburg from Buenos Aires. On board were 86 crew members, 53 of them young naval cadets learning how to sail this famous four-masted windjammer.

The *Pamir* was built by Blohm and Voss in 1905 and was originally powered by sails from its four masts. As a sailing ship it became famous in the early years of the 20th century in the "grain races" from Australia.

In the early 1930s the *Pamir* was bought by a Finn, Captain Gustav Erikson, and in 1951 changed owners again. In 1954 the ship was taken over by the Pamir-Passat Foundation. Although the *Pamir* kept its sails it also now had auxiliary oil engines.

At the time of its loss the *Pamir* was acting as a training ship for German naval cadets. It was 600 miles southwest of the Azores when it ran into a hurricane. The final message received from the ship said that the sails were ripped to shreds, the foremast had snapped off, and the ship was listing 45 degrees. That was the last that was ever heard of the *Pamir*. Of the 86 crew members on board, only six survived the ordeal.

Above: *Karl Otto Dummer, one of six survivors from the* Pamir, *greets his family on his safe arrival at Frankfurt.*

Left: *The four-masted German windjammer, the* Pamir.

LEWISHAM, ENGLAND

DECEMBER 4, 1957

Right: The aftermath of the collision at Lewisham. The list of casualties was all the greater because of the collapse of the bridge seen here on to passenger coaches.

Thick fog was a major factor in this incident which left 90 passengers dead and over 100 seriously injured. This rear-end collision involved a steam express traveling from Cannon Street station to Ramsgate in Kent and a local commuter train working between Charing Cross and Hayes.

The driver of the express ran through two warning signals without reducing speed and only began to decelerate when his fireman spotted a red light at close range. However, the express's speed was such that its driver, even though he applied the brakes, could not bring the train to a halt before it smashed into the stationary Charing Cross–Hayes train.

The 30mph (48km/h) impact was made more severe because the stationary train's driver had his brakes on full because he had come to a dead stop on an up gradient while waiting for a red signal to change once a train in front had moved off. The impact damaged many of

Left: One of the carriages from the Charing Cross to Hayes commuter train is removed from St John's station.

Right: The catastrophic damage inflicted on one of the passenger carriages involved in the crash.

the Hayes train's ten carriages. The eighth carriage was smashed when the ninth was propelled into it by the express's impact.

The front section of the express also suffered severe damage when its front carriage was thrown into the locomotive's tender by the force generated by the rapid deceleration. Both the carriage and tender were thrown sideways and smashed into the central pillar of a bridge carrying the line between Nunhead and Lewisham. Some of the bridge's girders crashed down on to the wreckage below, completing the destruction of the lead carriage of the express and damaging the second and part of the third carriages.

One of the express's carriages had to be cut up before it could be removed from the scene and the line was closed for more than a week. But for the quick thinking of its crew, a third train could have been involved in the accident. It was about to cross the damaged bridge when they brought it to a very timely emergency halt.

The fog, which had been blanketing the area for most of the second half of the day, was identified as one of the main causes of the accident. The two trains involved had been experiencing delays during the evening rush hour out of London: the Cannon Street–Ramsgate service was more than 60 minutes behind schedule when disaster struck and the Charing Cross–Hayes train was 30 minutes late. Investigators also highlighted problems with the signal system in the area of the accident. Installed in the 1920s, they were visible at close range from the cabs of diesel and electric trains, but were obscured by the boilers of steam locomotives at similar distances.

The driver of the express should perhaps have made more effort to check for a warning signal himself or asked his fireman to do so. He was twice tried for manslaughter. The first jury did not reach a decision and the second saw the prosecution offer no evidence because of the driver's mental condition. He was therefore acquitted.

Below: The warped remnants of the bridge that crashed down on to the train below at Lewisham. Huge amounts of steel fell when the bridge collapsed.

MUNICH AIRPORT, WEST GERMANY

FEBRUARY 6, 1958

This disaster achieved instant global notoriety, mainly because of the aircraft's prized passengers, the Manchester United Football Club's "Busby Babes" – a young and dynamic side that had just qualified for the semi-finals of the European Cup Championship in Belgrade, and was now returning home to Manchester via Munich.

The sunshine of Belgrade gave way to icy rain and snow as the British European Airways Airspeed Ambassador 2 closed in on Munich. As the aircraft descended through 18,000 feet (5400m) of cloud, the pilot activated the airframe anti-icing system, which heated the tailplane, fins, and wings to 60 degrees Celsius. Landing was normal except that huge plumes of slush sprayed up from the nose wheel. This would later prove to be critical.

A thick layer of slush enveloped Munich's runways that day, but at that time virtually nothing was known of the drag effects of a contaminated runway. The slush had now been measured at half an inch (1cm) on the busiest points of the runway, but its depth at the runway's end was still unknown.

After re-fuelling the pilots inspected the wings of the aircraft for ice and snow. Ice or snow on an aircraft's upper surfaces affects lift by impeding the airflow, but as there was only a slight film of ice that was thawing and running clear, the pilots deemed de-icing to be unnecessary. Later this became the basis for the authorities' charge of pilot negligence.

As the aircraft accelerated for take-off the co-pilot (flying as pilot) suddenly shouted "abandon take-off." The reason was "boost surging," caused by an over-rich fuel mixture, which was compounded by the thin air at Munich's high altitude.

After another abandoned take-off, this time due to a loss of power in one engine, the pilots conferred with the station engineer. One option was to re-tune the engines, but this was rejected since it would take all night. They decided upon one last attempt, this time releasing the throttles as slowly as possible.

However, when the flight crew eased off on the throttles to deal with the surging they found that they needed more runway to become airborne. The slush that was causing the trouble was at its deepest at precisely the point where the aircraft needed to achieve maximum acceleration.

With the nosewheel off the ground, and great plumes of slush rushing past the passenger windows, the flight crew committed themselves to take-off. But

Above: A BEA Airspeed Ambassador aircraft similar to the one that crashed at Munich.

Right: Rescue attempts at the scene of the crash were hampered by the icy rain and snow that enveloped the airport.

Below: A week after the crash salvage operators remove pieces of wreckage of the aircraft for investigation.

the aircraft never managed to reach the minimum speed required to become airborne. As the aircraft approached the end section of the runway, all the flight crew could do was watch as their speed plummeted. They ran out of runway well short of the speed needed to lift them off it.

The aircraft plowed through the snow beyond the runway, smashed through a wooden fence, and then squared up to a house and a tree across the road. In desperation the flight crew raised the undercarriage in an attempt to get airborne. But the left wing and the tail struck the house, and the tree tore apart the cockpit. Just 100 yards (90m) past the house the aircraft hit a wooden hut which sheared off the complete tail section. The remaining forward section catapulted across the snow for another 70 yards (64m) before coming to rest in a field.

Of the 44 people on board, 23 died, including eight "Busby Babes", four Manchester United staff, eight journalists covering the big match, a Yugoslav passenger, an air steward, and the co-pilot. The pilot survived, and after a decade of disputes between him and the German authorities, the drag effects of slush on take-off were finally properly tested and recognized. However, ice on the wings remained the official cause of the accident.

NEWARK BAY, NEW JERSEY, USA

SEPTEMBER 15, 1958

The exact reasons for this crash were never fully identified, as three of the individuals probably best able to shed light on the incident, members of the crew of commuter train 3314 from Bay Head junction, died in the accident. The destruction of Train 3314 took place on the steel bridge that spanned Newark Bay. Built in 1926 to replace the original bridge which dated back to 1864, it consisted of a pair of twin-track high-level bridges that could be raised to permit the passage of river traffic.

There were stringent safety procedures in operation to regulate the movement of trains and the raising and lowering of the central spans. An automatic warning signal had been placed some 1500 yards (1385m) from the bridge and a second stop warning some 200 yards (185m) from the lift. As a further safety measure, a derailer had been built some 25 yards (23m) beyond the second signal. Normally, shipping had priority and the warning signals were against the trains when the bridge was raised for their passage.

On September 15, a freighter called for the bridge to be raised as Train 3314, consisting of two diesel locomotives pulling five steel carriages, was due to leave Elizabethport station, the last halt on the line before Newark bridge.

The train left Elizabethport and headed for the bridge, its crew seemingly paying no attention to the two warning signals which showed that it was being raised to permit the passage of river freight. The train then hit the derailer at an estimated speed of more than 40mph (64km/h), careered along the sleepers for a little way, and then the locomotives and the first two carriages they were hauling plunged through the gap created by raising the bridge and sank into the murky waters of Newark Bay.

Right: *The second of the commuter train's three carriages to end up in Newark Bay is raised from its murky depths. A third coach was yet to be recovered from the scene when this photograph was taken.*

Remarkably there were no passengers in the first carriage, but most of those in the second were drowned. The train's third passenger carriage, initially caught on the bridge's pier, later plunged into the river. Most of the passengers trapped in this carriage escaped, rescued by several small craft that had hurried to the crash scene.

Investigators' reports suggested that the accident may have been initiated by the different safety procedures that applied to different sections of the track in the vicinity of the bridge. Some 12 miles (19km) south of the bridge, along a section of the New York and Long Branch, there was a system which enforced a 20mph (32km/h) speed restriction once a locomotive had passed a signal indicating caution.

However, from that point to the bridge, no such safety measure was in place. It was also discovered that the usual practice was for the bridge to be raised not to its full extent, but rather just sufficiently to allow for the height of the passing vessel. If it had been raised to its full height, concrete counterweights would have been lowered sufficiently to block the track and might have prevented Train 3314 from plummeting into Newark Bay. Although raising and lowering the bridge to its full height several times a day would have been time consuming, it could have reduced the casualties.

Above: A coach hangs from one of the bridge's piers with its forward portion hidden by the bay's waters.

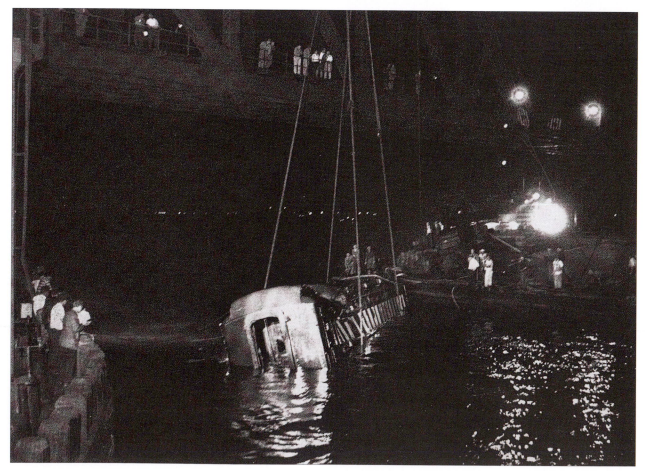

Left: A wrecked carriage is brought to the surface by a huge floating crane.

OUR LADY OF THE ANGELS SCHOOL, CHICAGO, USA

DECEMBER 1, 1958

Ninety-three people, most of them children, were killed when fire raced through a school in Chicago, USA, on the afternoon of December 1. The blaze was caused by an explosion in the boiler room: fanned by an icy wind, the flames swept rapidly through the brick building.

The fire broke out at Our Lady of the Angels school on the city's West Side only 18 minutes before classes were due to be dismissed for the afternoon. There were 1300 pupils and staff in the school at the time. Although most were led unhurt to safety, many children and several teachers were trapped at their desks where they burned to death or were overcome by smoke. The death toll was particularly high on the second floor – in one classroom there, rescuers later found the bodies of 22 children and two nuns. Another 100 or so people were taken to hospital: many of these were badly burned or suffering the effects of smoke inhalation, while others were injured after jumping from upstairs windows.

That there were not even more fatalities was due largely to the prompt action of staff and passers-by. As soon as screams were heard, priests and nuns dashed from the rectory next door with blankets and towels and tried to rescue those trapped inside the brick school building. Three men who happened to be walking past the school hurried in to help but, tragically, they were overcome by heat and were among those killed. The fire brigade was quickly at the scene, as were all available doctors and nurses, who treated many victims on the spot.

Right: School was almost over for the day when fire raced through the Our Lady of the Angels building, filling the upper floors with toxic fumes. Many pupils were saved by climbing down firemen's ladders, but 87 children and three nuns died in the blaze.

Right: After Chicago's worst school fire, firemen inspect one of the gutted corridors in the building.

For those who were not engulfed in flames, the main danger was from toxic fumes. Thomas Raymond, a 12-year-old pupil, later described how smoke had filled his upstairs classroom and the only way he could get fresh air was by lying on the floor. Then he broke the window by throwing books at it – he was preparing to jump when the fire brigade warned him not to. Thomas was one of many who clambered to safety down the firemen's ladder.

One brave nun singlehandedly saved 30 or 40 children from the fumes by making them crawl along the floor of the upstairs corridor and then rolling them down the stairs. Other children threw themselves onto the floor where the air was less smoky and kept still: most of these survived, although, by the time help arrived, many had been overcome by fumes and were pulled out unconscious.

As the fire still raged, the Chicago Fire Commissioner told the press "this could have been a touch-off" – American fireman's slang for arson. But subsequent investigations revealed that the disaster had been entirely accidental. Eleven-year-old Joseph Brocato told a doctor that he and a classmate had been emptying a wastepaper basket in the basement when they heard "a boom in the furnace room." Then the janitor ran out and shouted to them to get upstairs.

Right: An aerial view of the fire. Smoke pours from the building as firemen battle with the blaze. A crowd of over 10,000 people, many of them hysterical parents, had to be restrained by a police cordon. Of the 100 children brought out alive and taken to hospital, many were badly injured.

HANS HEDTOFT, ATLANTIC

JANUARY 30, 1959

The North Atlantic in winter is a treacherous place for shipping. The sea is icy, the weather can be stormy or foggy, and there is the ever-present threat of colliding with an iceberg. Even a ship specially built to withstand these conditions is not immune to disaster, as the fate of the *Hans Hedtoft* shows.

The *Hans Hedtoft* was built for Denmark's Royal Greenland Trading Company to sail between Denmark and Greenland. It was a specific requirement that the ship should provide a year-round service to Greenland, and that the sailings should continue even in the depths of winter, between January and March.

The builders tried to ensure that the vessel could survive the worst the North Atlantic could throw at it by designing the *Hans Hedtoft* with a double bottom, seven watertight compartments, and a reinforced bow. However, even these special features were not enough to protect the ship from the severity of the winter seas in the vicinity of Greenland.

The *Hans Hedtoft* began its maiden voyage on January 7, 1959, sailing out of Copenhagen for Godthaab. On January 29 the ship began the return leg of the voyage, under the command of Captain P. Rasmussen, with 55 passengers and a crew of 40.

On the 30th, the *Hans Hedtoft* faced heavy seas and gale force winds, but was able to make 12 knots. However, the ship collided with an iceberg shortly before 1200 hours some 35 miles (49km) south of Cape Farewell, on the southern tip of Greenland. Rasmussen immediately sent out a distress signal, which was picked up by a US Coast Guard vessel and a trawler.

The trawler raced for the stricken vessel's position but when it arrived it could find nothing. For several days after the loss, ships and aircraft crisscrossed the seas around the *Hans Hedtoft*'s last known position. Visibility was poor and no trace of the vessel could be found. On February 7, the search was called off.

The total loss of the *Hans Hedtoft* on its maiden voyage due to a collision with an iceberg was eerily reminiscent of the loss of the *Titanic*.

Left: The Danish ship Hans Hedtoft *sailing out of Copenhagen on its maiden voyage on January 7, 1959. The ship was destined never to enter Copenhagen harbor again – it sank without trace after colliding with an iceberg on its return voyage.*

EAST RIVER, NEW YORK, USA

FEBRUARY 3, 1959

Right: A rescue boat searching for victims of the Lockheed Electra crash. Of the 73 people on board 65 died when the aircraft failed to make an instrument landing at La Guardia airport at night and crashed into the East River.

Below: A giant derrick on a salvage boat lifts the tail section of the wrecked plane out of the river.

Among the eight survivors of this accident was an eight-year-old boy returning from Chicago with his mother, father, and two sisters. He was the only one of them to be rescued alive. Many of the victims survived the impact into the chilly waters of the East River, only to drown as they struggled to escape the rapidly sinking wreckage. Accident investigators were able to piece together an intricate picture of the errors leading to the crash from the accounts of the first and second officers, both of whom survived.

The Lockheed Electra airliner, belonging to American Airlines, was coming to the end of an internal flight from Chicago, Illinois, 730 miles (1174km) to the west. Although the flight crew had many years of air experience between them on a number of different aircraft types, they had only limited hours on the Electra, which had entered service with the airline less than two weeks earlier.

The weather conditions at La Guardia airport – the destination of the turboprop airliner – were poor, with visibility down to two miles (3km), low cloud base at only 350 feet (110m) and light rain and fog.

The conditions gave the flight crew little option but to make an approach based on instrument readings and the instrument landing system (ILS). This allows the aircraft to be guided to the ground by following a lateral heading beam and another angled vertical beam – both beams transmitted by ground-based navigational aids. The crew of the Electra were not familiar with an ILS landing at La Guardia, and as they approached the airport, they were under considerable pressure both to carry out the routine tasks of lowering the undercarriage and extending the flaps, and also to cope with the unfamiliar landing system.

There was little the flight crew could do to prevent the crash, which happened only seconds after the aircraft broke out of the dense overcast. The aircraft plowed into the icy East River, 5000 feet (1520m) away from Runway 22, where it should have landed. But for the presence of the tugboat *Dorothy McAllister* close by, it is doubtful whether anyone at all on the aircraft would have survived.

The US Civil Aeronautics Board, who investigated the crash, made serious criticisms of the captain and his two officers, and also airline operating practices. Instruments salvaged from the wreckage – much of which was recovered from the depths of the East River – showed potentially fatal anomalies. The altimeter had been inaccurately set at Chicago, and was reading nearly 120 feet (36m) too high. Added to this, the first officer, whose responsibilities included making regular reports to the captain on the aircraft's speed and altitude, was distracted by the complications and unfamiliar procedures of the instrument landing system at La Guardia.

These two errors on the part of the flight crew, combined with their unfamiliarity with the aircraft, were deemed to have caused the crash. Evidence given to the Civil Aeronautics Board by the two junior flight officers indicated that the design of the instruments confused them; this may have led to mistaken instrument readings. In addition, there was inadequate approach lighting to Runway 22, and the rain and poor visibility may have combined to give the pilot the impression that he was higher than he really was.

Left: A marine crane lifts wreckage from the Electra onto a barge. Many of the bodies of the victims of the disaster were never recovered from the river.

The 1960s saw an explosion in cheap mass travel, both at sea and in the air. Against a background of growing prosperity in which the disposable incomes of ordinary people grew steadily, various companies sought to tap the market for inexpensive holidays. Perhaps inevitably, demand somewhat exceeded supply, forcing operators to cut corners or rely on somewhat antiquated technology. A case in point was the cruise liner *Lakonia,* which sank in December 1963 with the loss of more than 130 lives. Although fire eventually destroyed the ship, it was clearly in the twilight of its career, having been built in 1931. The *Yarmouth Castle*, which sank two years later, was even more ancient, having been launched in 1927. Similar disasters were also befalling aircraft. Although more modern types were increasingly entering service, they could still fall victim to the weather, human error, or mechanical failure.

Although such holiday-related incidents were growing steadily, the greater majority of reported catastrophes revolved around work and industry. Train commuters were still being killed and locomotives and rolling stock wrecked due to collisions or derailments, while cargo vessels were succumbing to rough seas, fire, or faulty navigation. New technologies were always being introduced to minimize these risks, but as ever they were neither universal nor foolproof, and could not guarantee 100 percent safety.

Background: The Torrey Canyon, *one of a new breed of supertankers for the bulk transport of oil, gradually sinks beneath the English Channel in March 1967.*

1960-1969

NEW YORK, USA

DECEMBER 16, 1960

Four years after the collision of a United Airlines Douglas DC-7 and a TWA Super Constellation over the Grand Canyon, a similar accident occurred over the district of Brooklyn in New York. It involved the same two carriers, virtually identical aircraft, and it cost the same number of lives. It also showed that, despite a program of improvements, US air traffic control procedures were still fraught with problems.

The United Airlines DC-8 was on approach to New York International airport at the end of a scheduled service from Chicago. Weather conditions were poor, with visibility reduced to the extent that the crew were forced to rely on instrument readings to fly the aircraft.

The TWA Lockheed Super Constellation had originated at Dayton, Ohio, and was to land at La Guardia, New York's other major airport. At 1033 hours on December 16, 1960, as the two aircraft were descend-

Below: The tailplane and wing of the crashed Douglas airliner.

Right: The impact of the exploding Douglas DC-8 wrought havoc in the Park Slope area of Brooklyn.

ing toward their separate destinations, they collided in cloud over Miller Army Air Field on Staten Island.

The United Airlines DC-8 crashed into the upper fuselage of the Constellation from behind, causing the latter aircraft to break into three parts and plummet to the ground on Staten Island from a height of 5000 feet (1500m). The crippled DC-8 carried on for another 8.5 miles (13.5km) before impacting in the Park Slope area of Brooklyn.

The resulting explosion caused extensive damage to buildings in the vicinity, and killed six people on the ground. An 11-year-old boy survived the crash of the jetliner but died of his injuries in hospital soon after, joining the list of 134 fatalities.

Left: Pieces of wreckage from the DC-8 are loaded aboard a truck during the clearing-up operations following the aircraft's horrific crash into the streets of Brooklyn.

DARA, PERSIAN GULF

APRIL 8, 1961

Ships can be tempting targets for terrorists. When an explosion – followed by fire – sank the *Dara* in the Persian Gulf there was no thought initially that this was due to sabotage. But when divers went down to examine the wreck they found evidence of a bomb. The device had exploded in a passageway just above the *Dara*'s engine room, sending out a large fireball that engulfed parts of the ship.

The final voyage of the *Dara* began on March 23. The ship's master, Captain Elson, had 600 passengers on board. The outward leg to Basra went without a hitch and Elson then began the return trip, stopping at Korramshahr, Kuwait, and Bahrain before reaching Dubai on April 7. Elson took on more passengers and then decided to leave harbor early. The weather was worsening – there were gale-force winds and heavy hailstorms – and the vessel would be safer at sea. By the early morning of April 8 the weather had improved and the *Dara* turned back to Dubai.

At 0445 hours the *Dara* was rocked by an explosion, which stopped the engines and put the steering out of action. Frightened passengers and crew began to abandon ship – as the flames spread rapidly, many people threw themselves into the Gulf waters. A converted landing craft, the *Empire Guillemot*, saw the flames, sent out a distress signal (the *Dara*'s communication system had been put out of action), and rushed to the *Dara*'s aid.

Three British frigates also hastened to help the stricken ship, and managed to put out the flames.

Although more than 580 people were rescued, 241 others died in the disaster. An attempt was made to tow the *Dara* to port, but the ship sank on April 10.

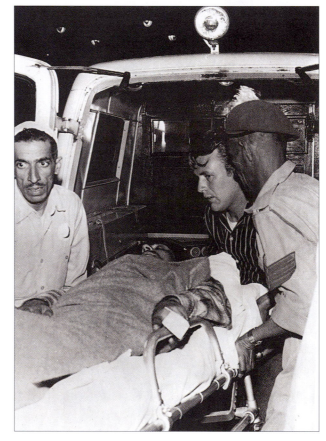

Left: A survivor from the Dara *is put into an ambulance after being brought ashore from a rescue ship.*

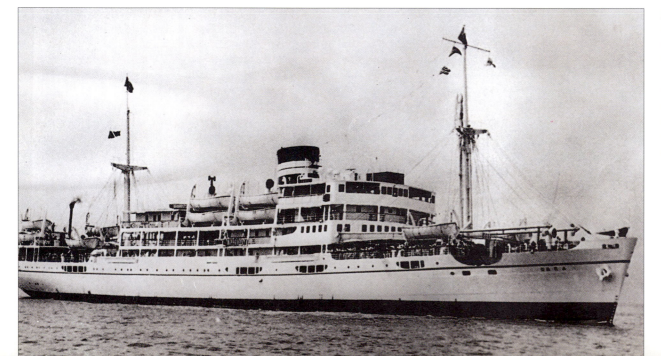

Left: The liner Dara, *owned by the British India Steam Navigation Company, was in the Persian Gulf when an explosion rocked the ship, putting it out of action.*

HARMELEN, NETHERLANDS

JANUARY 8, 1962

This mid-winter incident involved an express traveling between Utrecht and Rotterdam and a local commuter train working the eastbound track between Rotterdam and Amsterdam. Both trains were moving at high speed when the impact occurred and there was fog about. The incident took place just as the stopper was approaching the junction at Harmelen, a little to the east of Woerden. At this point, the track layout was such that trains traveling between Woerden and Amsterdam had to use a section of westbound track for a distance of approximately 25 yards (23m).

The head-on collision occurred on this short section of track. The express, which was carrying 900 commuters on their way back to work after the weekend break, smashed into the stopping train with considerable impact. Three of the stopper's coaches were destroyed, while six of the express's 11 coaches were also badly damaged. The carnage caused by the smash was such that rescue workers toiled throughout the day to sift through the twisted wreckage and help the dazed and injured survivors.

Responsibility for the crash was laid at the feet of the express's driver. He had missed one signal notifying him of the train ahead and only began to apply the brakes when he saw a stop signal registering danger. He was able to reduce speed from nearly 80mph (128km/h) to 65mph (104km/h), but could not prevent the collision in the stopping distance available to him. With 93 dead, this incident was one of the worst ever collisions experienced on the Dutch railroad network. Subsequent recommendations included the introduction of more modern safety devices.

Above: A Dutch doctor (dressed in white coat) prepares to enter the remains of a passenger coach to give medical assistance to those injured at Harmelen.

LAKONIA, ATLANTIC
DECEMBER 22, 1963

For some of the 651 passengers and 385 crew who embarked for a Christmas cruise aboard the Greek Line's 20,314-gross ton *Lakonia* on December 19, 1963, the voyage was to be the most memorable of their lives. But for more than 130 of them, it was destined to be their last voyage. The *Lakonia* set sail from Southampton on the south coast of England at the start of an 11-day cruise to the Canary Islands.

The *Lakonia* was not a new ship. It had been completed by Nederlandsche Shipbuilding in 1931 for the Nederland Royal Mail Line. The company intended that their new liner, called the *Johan van Oldenbarnevelt*, should serve the Dutch East Indies from Amsterdam. World War II intervened and the ship served as a troop transport, but returned to its old route in 1946. It later sailed to Australia packed with migrants, underwent a number of refits, and then trans-

ferred to the lucrative round-the-world cruise business. In 1962 it was sold to the Greek Line, renamed the *Lakonia*, and began serving the Southampton–Canary Islands route in spring 1963.

The Christmas cruise began to go wrong on December 22 when the *Lakonia* was some 180 miles (288km) north of the island of Madeira. A fire started in the ship's hairdressing salon. A large bang followed, possibly caused by pressurized containers in the salon exploding, and thick black smoke began to spread throughout the vessel. In the confusion many passengers began to panic.

The ship's crew, operating under the command of Captain M. Zarbis, gave orders that life-preservers should be put on, but that was as far as the emergency instructions went. There seems to have been a problem with the ship's public address system. Instructions given by word of mouth to passengers were conflicting. Some were told to stay below in their cabins,

Above: Smoke pours from the stricken liner Lakonia *as it is engulfed by fire about 180 miles (288km) north of Madeira.*

164

Right: Even though the Norwegian tug Herkules attempted to tow the burning Lakonia to port, the liner was too badly damaged to save, and sank soon after this picture was taken.

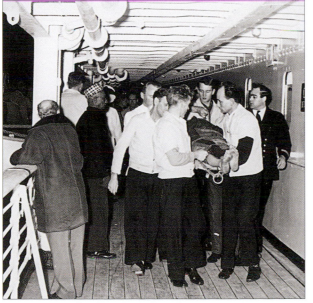

Right: An injured passenger from the Lakonia is lifted on board one of the rescue ships after being picked up by one of the lifeboats.

others to go to the dining room. Passengers who managed to climb up to the boat deck found a scene of confusion – the crew seemed to be having considerable difficulty in launching the available lifeboats.

The *Lakonia* was able to send out a distress call that was heeded by a number of vessels sailing in the vicinity. The first to arrive was the *Salta*, an Argentinian passenger liner. The *Salta* was joined by four other vessels – the *Centaur, Charlesville, Export Aide,* and *Montcalm.* Working in unison, the rescuers took on board over 900 people from the *Lakonia.* Nearly 90 people were known to have died in the fire and a further 42 were unaccounted for but presumed dead.

The *Lakonia* itself was too far gone to save. The fire had taken an irreversible hold on the ship. A Norwegian salvage tug, the *Herkules,* was able to get a line out to the *Lakonia* on Christmas Eve and began to head for the nearest port. However, the damage to the *Lakonia* was far too great for it to stay afloat and the liner succumbed to the Atlantic on December 29.

CUSTOIAS, PORTUGAL

JULY 26, 1964

Below: One complete side of this carriage was torn away by the violence of the impact with the stone piers of a bridge.

There remain several questions to be answered about the direct cause of this crash which produced close to 100 fatalities and left 79 people seriously injured. The specifics of the incident are not in doubt, however. A weekend train was carrying large numbers of Portuguese day-trippers who were returning to their homes in Oporto after a pleasant day out at a local beach resort. The train was heading for the city's Trindade station and was grossly overloaded, carrying nearly three times its recommended number of passengers. Many passengers had pushed their way on to the train rather than wait for the next service to depart.

As it approached a curved section of track some three miles (4.8km) outside the city, the first of the diesel-powered railcar's coaches was derailed and the second then smashed into the stone supports of a viaduct that crossed the line at that point. Fortunately, a third coach was brought to a halt.

Rescue workers faced a difficult task in searching through the train and administering aid to the many casualties. They had to work at night in the confines of a steep-sided, narrow-bottomed cutting filled with wreckage. Vital equipment had to be lowered down the slopes by rope and the injured had to be carried out using improvised stretchers or were placed on the backs of the emergency services' staff for the painful haul to the top of the cutting.

Many of the injuries were so severe that there was a real danger that the supplies of blood available for transfusions might run out. A hurried call for fresh donors was answered by hundreds of people in the vicinity of the crash site.

Investigators later discovered that the train approached the curve at a speed of more than 60mph (96km/h), considerably in excess of the recommended safety limit on the section of track, and it was this that brought about the derailment. The local media reported that the accident had been caused by the parting of the train's couplings. Whether this was the case or not remains difficult to establish.

Normally, a sudden uncoupling will halt both halves of the train automatically. The rear part may have the momentum to nudge into the forward section, but not usually with sufficient momentum to cause anything worse than minor injuries. Alternatively, if a rear carriage is derailed, the lack of alignment between it and the carriage in front is usually sufficient to bring about a coupling failure.

Equally, this can lead to sideways movements in a carriage, making it highly liable to collision with track-side obstacles, such as a signal gantry, the walls of a cutting, or the supports of a bridge. This argument, if applied to the crash at Custoias, suggests that a derailment led to the uncoupling and not an uncoupling which led to the derailment.

Right: Rescuers mill around one of the coaches involved in the accident at Custoias. The train was packed with trippers returning home after a day at the seaside.

YARMOUTH CASTLE, CARIBBEAN
NOVEMBER 13, 1965

The *Yarmouth Castle* had a long, if not particularly distinguished, career. Launched in 1927 as the *Evangeline*, the liner served with the US Navy during World War II. It was bought by its final owners, Yarmouth Cruise Lines, in 1963. The owners renamed the ship, but the new name could not revitalize the aging vessel, which was plagued by engine problems. Yarmouth Lines decided to switch the vessel to the cruise route from Miami to Nassau in the Bahamas.

In November 1965, the *Yarmouth Castle*, with 372 passengers and 174 crew members on board, set sail for Nassau. At around 0035 hours on the night of November 13 a fire broke out in cabin 610 and quickly spread along the ship's corridors to the decks above. The ship's master, Captain Byron Voutsinas, was not told of the fire for 25 minutes. He investigated the blaze and returned to the bridge, but failed to send out a distress signal. In fact, the fire was so out of control that he had to give the signal to abandon ship some 20 minutes later. By this time the flames had engulfed the bridge, and a large number of lifeboats. The ship's radio room was also ablaze, so that distress signals could not be sent out.

Two ships, the *Finnpulp* and the *Bahama Star*, did spot the burning *Yarmouth Castle* and raced to its aid. The first lifeboat to be picked up included Captain Voutsinas and some of his officers. Many of the ship's passengers had to be rescued from the sea. The two rescue ships eventually saved over 450 passengers and crew, but there was no saving the *Yarmouth Castle*. The ship, ablaze from bow to stern, heeled over in the early morning and sank.

Above: This dramatic picture of the Yarmouth Castle *ablaze was taken by a passenger on the* Bahama Star, *one of the ships that went to the rescue. An unlaunched lifeboat can be seen on the left.*

MOUNT FUJIAMA, JAPAN

MARCH 5, 1966

Prior to this incident, British Overseas Airways Corporation (BOAC) had suffered no fatal air accidents during the 1960s, an admirable record considering the vast increase in the volume of passengers carried by the airline. This record was broken tragically on March 5, 1966. It was on this date that a Boeing 707 operated by the company was torn apart in the air by violent turbulence at an altitude of 15,000 feet (5000m), near the towering summit of Mount Fujiama in Japan.

Captain Bernard Dobson had taken off from Haneda airport (serving Tokyo) for the 1100-mile (1800-km) leg to Hong Kong, which was part of a scheduled round-the-world service that had left London four days previously. In order to give his (mostly American) passengers a view of the mountain, he deviated from the designated airway and flew west.

Fifty miles (80km) south west of Tokyo, the aircraft flew into severe turbulence. Buffeted by gusts of wind, it began to disintegrate. Parts of the aircraft were detached and fell off. The rear control surfaces and the rear fuselage were torn off and the stricken jetliner began to tumble out of the sky. As it fell, the fuselage and main engine assemblies were ripped off.

Pieces of wreckage and the bodies of all 124 people on board were recovered from a wide area at the foot of the mountain.

This crash shook the aviation community, more particularly as it occurred less than 24 hours after the crash of a Canadian Pacific DC-8 at Haneda, with the loss of 64 lives.

Below: Japanese soldiers searching for bodies in the wreckage of the crashed BOAC Boeing 707.

HANSEATIC, NEW YORK

SEPTEMBER 7, 1966

As the loss of the *Hanseatic* proves, a ship does not have to be on the high seas to suffer a major catastrophe. The *Hanseatic* was lying peacefully at anchor in New York harbor when the failure of a small component in the engine room led to a fire that swiftly engulfed the ship. It was the end of a 30,000-ton vessel that had, as a troopship, survived the horrors of World War II.

The *Hanseatic* began life as the *Empress of Japan*, plying between Vancouver and Yokohama in Japan. War brought an understandable change of name to the *Empress of Scotland* in 1942. Released back to civilian use in 1948, the ship was eventually bought by the Hamburg-Atlantic Line early in 1958 and renamed the *Hanseatic*. The *Hanseatic*'s main route was between Germany and New York but it was also employed as a

winter cruise ship. The vessel was destroyed shortly before it was due to set out on one of these cruises on that fateful September 7. Only three of the scheduled 425 passengers were on board the *Hanseatic* when fire broke out at about 0730 hours. If the fire had started later, nearer the 1130 hours sailing time, more passengers would have been on board and the loss of life might have been considerable.

The fire started in the engine room and was caused by either a broken gasket or a faulty fuel line. Whatever the cause, the flames took hold rapidly, spreading into two more engine rooms and making their way undetected via vents to the passenger decks. The fire was eventually brought under control, but the damage was too extensive to make repair an economically viable option. The *Hanseatic* was finally towed to Hamburg and sold for scrap in December 1966.

Above: The Hanseatic on fire in New York harbor. The 500 crew members and the three passengers aboard were all evacuated safely.

Right: The German liner Hanseatic setting sail from Southampton.

HERAKLION, AEGEAN SEA

DECEMBER 12, 1966

Above: The Greek car ferry Heraklion *which sank midway between Crete and Piraeus with the loss of over 200 lives.*

Lax discipline and flouting of safety regulations on passenger ferry services generally only come to light after a tragedy has occurred. Such was the case with the Greek ferry *Heraklion*, which foundered in rough seas on its regular trip from Crete to Piraeus. A rescue vessel reached the site within 30 minutes but no trace of the ship was found. There was some debris and huge quantities of fruit floating on the rough seas, but no sign of the *Heraklion*. Later, 47 passengers and crew were found on the island of Falconcra but the remaining 231 were presumed lost. What had happened?

The 8,922-gross ton *Heraklion* had begun life as the *Leicestershire*, which sailed the British–Burma route, but was sold to Greece's Typaldos Lines in 1964 to sail between Piraeus and Crete. On December 7, 1966, the weather on the Aegean Sea was treacherous and stormy. Nevertheless, the *Heraklion* sailed out on the normal ferry service from Crete. The buffeting the ship took caused considerable problems on the cargo deck, loosening the ties that held cars and lorries. One 16-ton trailer broke loose and smashed into the cargo door, which opened the vessel to the raging waters outside. The *Heraklion* flooded rapidly. The vessel sent a distress signal at 0200 hours on December 8, and the Greek air force and navy, as well as two British warships, responded quickly. But their arrival was too late for most on board the *Heraklion*.

The loss of the ferry was investigated and the Typaldos Lines was severely criticized. The board reported that the ship did not have an established emergency drill, the SOS was sent out too late, and the ship's officers had failed in their duty. Charges of manslaughter and forgery were brought. Two of the company's senior officials were given jail sentences.

TORREY CANYON, ENGLISH CHANNEL

MARCH 18, 1967

Right: The super-tanker Torrey Canyon *breaking up after it ran aground on Seven Stones Rocks.*

The grounding of the *Torrey Canyon* super-tanker on Pollard Rock, the most western of the Seven Stones Rocks off England's Land's End, heralded a radical rethink in the dangers and methods of coping with oil spills at sea. When the calculations were made, it was estimated that around 100,000 tons of crude oil had leaked into the waters off England and France. This oil spill was 10 times greater than any previous spillage. Governments and anti-pollution agencies did not have any experience of dealing with an environmental catastrophe of this magnitude.

The 118,285-ton *Torrey Canyon* was sailing from the Persian Gulf for the Milford Haven oil terminal in south Wales. The tanker, captained by the experienced Pastrengo Rugiati, was sailing under a Liberian flag.

Although owned by the Barracuda Tanker Company of Bermuda, on its fateful voyage the *Torrey Canyon* was chartered for British Petroleum. Captain Rugiati was under a little pressure from the Milford Haven authorities. If he did not make the evening tide on March 18, it was unlikely that he and his valuable cargo would be able to dock until the 24th.

The incident began at 0630 hours, when one of the vessel's officers picked up a radar reflection off to starboard. He was expecting an echo off the Scilly Isles, but from port. He ordered a change in direction to take the vessel to the west of the radar echo, but Rugiati intervened, taking the *Torrey Canyon* back on to its original heading and – unusually – puting the ship on autopilot. A nearby lightship saw the danger and fired warning rockets but to no avail. The tanker hit Pollard Rock at 0915 hours. Rugiati ordered "full astern" but the ship was stuck fast. The sound of metal grinding on rock also suggested that its bottom was being torn out.

Over the following days efforts were made to pull the *Torrey Canyon* off Pollard Rock. None was successful, and after an explosion in the engine room left one crewman dead, the vessel was abandoned. As the weather worsened, oil, which had only been seeping out of the damaged hull, began to pour out after the vessel broke its back on March 27. Aerial photographs revealed an oil slick some 35 miles (56km) long and up to 15 miles (24km) wide. Beaches in southwest England, Brittany, and the Channel Islands were coated in oil. Tourism was badly damaged and thousands of seabirds died. The fishing industry was also severely disrupted.

The authorities tried several methods to minimize the damage. Detergent was sprayed to break up the slick and booms were deployed to contain it. Finally, the Royal Air Force bombed the *Torrey Canyon*, hoping to set fire to the oil remaining on board. None of these measures was wholly successful, although the first bombing mission did ignite some oil, which burned for two hours. The subsequent Liberian board of inquiry concluded that "the master [Rugiati] alone is responsible for this casualty." It recommended that Rugiati's captain's license be revoked because of his negligence and the severity of the incident.

Left: A helicopter lowers a large compressor on to the deck of the Torrey Canyon *as salvage crew work desperately in an attempt to refloat the tanker on the next high tide.*

Below: After the crew of the Torrey Canyon *had been evacuated, the RAF set fire to the oil on board, which burned for two hours, sending out clouds of dense black smoke.*

L'INNOVATION, BRUSSELS, BELGIUM

MAY 22, 1967

Three hundred and twenty-two people were killed in a fire which almost completely destroyed L'Innovation, a large department store in the Rue Neuve in the downtown district of Brussels, Belgium.

The fire started in the furniture department on the fourth floor during the lunchtime rush hour. The shop would have been busy with shopping crowds at this time on any day, but today it was particularly full with visitors to the first day of "American Fortnight," a huge

Left: *Despite the strenuous efforts of the firefighting forces, the largest department store in Brussels, L'Innovation, was almost entirely destroyed by a fire that started on the fourth floor during the busy lunchtime shopping period.*

Far right: When the smoke finally cleared from the ruins of the department store, only a heap of charred rubble remained.

in-store exhibition of US merchandise. There are thought to have been at least 1000 people on the premises at the time of the outbreak. To mark the occasion, much of the inside of the shop had been specially decorated with Stars and Stripes flags – these added fuel to the blaze, as too did a large stock of paper dresses specially imported from the US.

The alarm system did not go off – probably because the fire had knocked out the electrical systems – and thus many shoppers remained unaware of the danger they were in even after flames became readily visible from the outside of the building. Although an alternative signal was given, most customers failed to understand the significance of repeated ringing of the service bell.

The fire quickly gained a strong hold and violent explosions shook L'Innovation every few minutes. The blaze raged out of control and soon threatened to spread to a large number of adjacent buildings on the block. One whole side of the store collapsed with a roar, covering fire engines in rubble but miraculously injuring none of the firemen.

However, Deputy Fire Chief Jacques Mesmans broke both his legs after picking up a woman and jumping with her in his arms out of a second floor window. Many of those trapped high up the building

Below: As a pall of black smoke rose over the blazing building, many shoppers leaped to their deaths from upper floor windows, their clothes in flames.

faced the desperate choice between jumping and burning. Three people were killed by leaping from the roof, where they had been trapped by flames. Meanwhile, many others perished horribly as they were transformed into living torches on window ledges in full view of the world.

As the first people were led out through the main exit, one man rushed in and started looting the store. He was quickly overpowered and taken away by police.

After the fire, L'Innovation and two surrounding blocks had to be demolished: they had been due for redevelopment anyway. There was speculation at the time that the fire may have been started deliberately by anti-American demonstrators, but this was never proved – a likelier cause seems to have been an electrical fault or a carelessly discarded cigarette butt.

MEDITERRANEAN SEA, OFF TURKEY

OCTOBER 12, 1967

When an airliner explodes in the air with the loss of all on board, it can often be difficult to establish the cause of the explosion. This was the case with the British European Airways De Havilland Comet 4B.

British European Airways was one of the first operators of the De Havilland Comet 4B, which it first put into service in May 1952.

In the darkness of early morning on October 12, 1967, one of their Comet fleet took off from London Heathrow airport on a flight to Athens, Greece. Having completed the first leg, the aircraft left Athens for Nicosia on the island of Cyprus, about 559 miles (900km) to the west. The last communication with the aircraft was at 0718 hours, when it was flying on a westbound heading at a height of 29,000 feet (8839m).

At around 0730 hours the Comet was crippled by an explosion. Out of control, the aircraft tumbled 14,000 feet (5,000m) before the fuselage broke into two parts. The wreckage hit the water of the western Mediterranean about 100 miles (150km) east south east of Rhodes. The bodies of 59 of the 66 passengers and crew were recovered, but the depth of water in the region hampered the operation to salvage the wreckage.

At first it was thought the Comet had suffered a structural failure, but when one of the seat cushions was recovered from the water and analyzed it was found to contain traces of an unidentified high explosive. Terrorist activity was immediately suspected, particularly in view of the continuing conflict over the sovereignty of Cyprus, in which Britain was taking an active military role.

One theory suggested that a bomb had been placed on board in an attempt to assassinate the leader of the Greek forces in Cyprus, after he had mistakenly been identified as one of the passengers. Another theory was that the bombing was part of an insurance fraud – at least two of the passengers had unusually high cover. But no firm evidence for either theory was ever forthcoming – and the cause of the explosion has never been fully established, nor the killers apprehended.

Below: A BEA de Havilland Comet 4B airliner similar to the one that crashed in the Mediterranean on October 12, 1967, with the loss of all lives.

HITHER GREEN, ENGLAND

NOVEMBER 5, 1967

Left: A shattered coach, the one in which many of the fatalities occurred, is secured to a transporter before being removed from the site of the Hither Green accident.

This derailment took place because of a small section of broken rail on the main line running from the town of Hastings in East Sussex to London's Charing Cross station. The chief crash investigator concluded that the amount of maintenance carried out on the track was considerably less than was warranted on a busy line and contributed to the failure of the rail. On the Sunday evening in question, a packed train approached Hither Green, south of London, at 70mph (112km/h) and was braking to meet the 60mph (96km/h) restriction that began at the station.

As the train traveled over the fractured joint between two rails, only the front axle of the third carriage was derailed and the remainder of the service stayed on the track for a further 450 yards (415m). Here, however, much of the train that had not been effected by the

initial problem was derailed when the original derailed coach reached a section of crossover track.

Many of the previously unaffected carriages were dragged off the track by the impact of the original derailed carriage with the crossover rails. Only the leading car remained on the track and four of the train's 11 units were thrown on to their sides. The list of casualties consisted of 49 dead and 78 injured, nearly 30 of these seriously.

The damage to the rail had occurred on a frequently busy section which was undergoing up-grading work. The work was seemingly progressing well and four months before the accident the speed limit in the area had been increased to 95mph (152km/h) for certain trains with more advanced bogies.

However, trains of the Hastings type were restricted to a maximum of 75mph (120km/h) because they had been modified by having their bogies made less flexible

to reduce the amount of swaying that took place at higher speeds and could prove dangerous in narrow tunnels.

Careful analysis of the fractured track and the surrounding area quickly suggested how the break could have happened. It was found to be located at a point where two short rails had been set between lengths of welded track. The bed of ballast which was supposed to support the two short rails was found to be insufficient to prevent movement particularly at the junction between the two and the passing of trains had led to the evolution of severe stress fractures.

These gave way, fracturing a triangular-shaped piece of rail and splitting a plate holding the two short rails together, as the Hastings train passed. To minimize the risk of similar accidents, the installation of continuously welded track was speeded up and the use of the short rails prohibited. These improvements came too late for the many casualties on the Hastings–Charing Cross train and their families.

Above: A carriage lies on its side following its violent derailment. The roof has been ripped off by the force of the accident.

Right: Working amid scattered wreckage and twisted track, heavy cranes begin the task of removing one of the train's smashed carriages.

WAHINE, NEW ZEALAND

APRIL 11, 1968

A shipwreck can have profound repercussions for its passengers, crew, the vessel itself, and its owners. The loss of the 8948-gross ton *Wahine* in April 1968 left more than 50 people dead or missing. The ship was fit for nothing but scrap, and the disaster sparked a chain of events that led to the closure of the ship's parent company, the Union Steamship Company of New Zealand. This venerable company, which had been founded in 1875, had its reputation ruined by the incident.

The *Wahine*, with room for more than 900 passengers, was a hard-working vessel. Built in Glasgow in 1966 by Fairfield Limited, it was earmarked to carry passengers and cars between Wellington on New Zealand's North Island and Lyttelton on the South Island. It was expected to make six overnight crossings between the two ports each week. The seas around New Zealand were well known for ferocious storms.

On April 11 the *Wahine* was making its way to Wellington in the teeth of a ferocious gale with winds gusting at over 120mph (192km/h). There was zero visibility. The ship struggled to maintain its course, but its captain and helmsman found it increasingly difficult to keep control of the *Wahine*. Shortly after 0630 hours the vessel was thrown on to the sharp rocks of Barretts Reef close to the entrance to Wellington harbor. Somehow the *Wahine* broke free of the reef but the damage had been done – the hull was holed and the starboard propeller had been severed.

Reports from the engine room indicated that water was flooding in. If the vessel could not reach a safe

Below: The Wahine at midday on April 11. The ship has been damaged and is listing to starboard, and the first lifeboats are being launched.

anchorage, it would undoubtedly founder. During the attempt to find safer waters, the *Wahine* was again thrown on to rocks at the entrance to a channel known as Chaffers Passage. There was no saving the vessel. Water poured through the twisted and torn hull plates, causing the *Wahine* to heel over to starboard.

At 1330 hours the order to abandon ship was given. As the *Wahine*'s port-side lifeboats were unusable because of the list, those on board – more than 700 passengers and crew – had to take to the lifeboats on the starboard side. Reports indicate that there was some panic but matters eased when the violent winds subsided. Although the seas were still rough, the evacuation was carried out with a degree of control for the most part. Nevertheless, 50 lives were lost.

What of the vessel itself? The *Wahine* eventually rolled over completely on to its starboard side and was declared a total wreck. Repair costs were far too high to make restoring the ship worthwhile. The *Wahine* was righted and refloated, but then sold for scrap.

The Union Steamship Company finally got around to replacing the *Wahine* in 1972, but failed to recognize the change in public opinion that followed the disaster. New Zealanders turned to other forms of transport to commute between Lyttelton and Wellington. In 1974 the Union Steamship Company folded.

Left: The day after the disaster the capsized Wahine *lies on its side in Wellington harbor.*

Right: Survivors from the Wahine *are brought ashore in a lifeboat. Over 50 people were lost in the disaster.*

FARMINGTON, WEST VIRGINIA, USA

NOVEMBER 20, 1968

Seventy-eight men died after a series of methane gas explosions sparked terrible fires below ground in the Consol No. 9 coal mine near Farmington, 10 miles (16 km) outside Monongay, West Virginia, USA.

The shallow mine extended eight miles (13 km) from east to west and six miles (9.5 km) from north to south through an underground coal seam between the small towns of Farmington and Mannington. Below the excavations lay a rich but still untapped reservoir of oil and natural gas.

At about 0540 hours on a winter's morning, one large underground explosion and three secondary blasts sent massive shock waves through the mine and up to the surface. The most seriously damaged building at the pit head was the lamphouse, a store room in which were kept not only the miners' lamps but also the one complete list of the names of those on shift at the time. Of the men who had gone to work 600 feet (180m) underground at midnight, 21 were quickly hauled to safety. The rest, however, remained unidentified until frantic calls to their homes enabled a new register to be compiled. By 0800 hours, when the shift was due to end, it had been established that there had been 99 men in the mine, and that 78 of them were unaccounted for.

Hours after the main explosion, dense black smoke from the fire still rose in a column above one of the ten main adits (entrances) to the mine. Teams of rescuers stood ready at the other nine entrances but were not allowed to go down at once for fear of further explosions. During the day engineers put bricks and concrete over two of the ventilation shafts in an attempt to direct the flow of air away from the fire and towards the area in which it was thought most likely that any survivors might have taken refuge.

But at 2200 hours that evening, another blast of methane gas blew out the seals and flames again roared through the wrecked entrances to the mine. Rescue workers now began to think it would be days or even weeks before they could go down. The only hope for the trapped men was if they had managed to

Left: Clouds of smoke billow from the Llewellyn Portal of the Consol No. 9 coal mine on November 20, 1968, after a series of massive explosions started fires in the mine that were to trap and kill 78 miners.

Right: A miner tests the air for gas. It was thought to be explosions of methane gas that started the disastrous fires that swept through the Consol mine.

escape into an area near where the 21 survivors had been found. But there were now only two working ventilators and it was doubtful if the men could have reached the areas these fans were supplying.

Then, during the night, there was a further explosion and the fire turned back towards where the men were trapped. More smoke belched out from the main shaft. The only way to stop the fire was to starve it completely of air, but to do so would also kill any

survivors. Ventilation holes were drilled into the ground directly above the chambers in which the men were thought to be. Sensors suggested that this part of the mine was high in methane and low in oxygen, but could still have supported life.

Nevertheless, no contact with the trapped miners was ever made. On November 30, hope was abandoned and the mine was finally sealed with its 78 victims entombed for ever.

The period between 1970 and 1979 was characterized by a small but growing awareness of the fragility of the environment, a factor in part brought about by the growing number of major incidents that were responsible for significant ecological disasters. Pollution, which had never before featured to any great extent in the reporting of incidents, came to the forefront regularly. Chief among such events was the sinking of the supertanker *Atlantic Empress* in the Caribbean following a collision with another supertanker, the *Aegean Captain*, in July 1979. On land, the chemical plant at Flixborough, England, was devastated by fires caused by an explosion that tore through the facility in June 1974.

A second feature of the period was the number of incidents that occurred in places of entertainment or leisure. More often than not younger people were the victims. The Club Cinq-Sept in Grenoble, France, was destroyed by a lighted match that created an inferno in November 1970 and led to the deaths of some 150 young adults. Equally, the Summerland Leisure Center in the Isle of Man was badly damaged by fire sparked by an electrical fault on August 2, 1973.

Yet despite these tragedies, notable though they were at the time, other events were even more shocking. Not least among them was the March 1977 air crash in the Canary Islands that left 583 people dead.

Background: *The final moments of one of the finest ocean liners, the* Queen Elizabeth, *in Hong Kong harbor during 1972. The vessel, renamed* Seawise University, *fell victim to fire.*

1970-1979

CLUB CINQ-SEPT, GRENOBLE, FRANCE

NOVEMBER 1, 1970

Left: The ruins of the Cinq-Sept club after the fire in which 146 young people died.

Ablaze which began accidentally when a lighted match fell onto a plastic foam cushion ended shortly afterwards in the death of 146 young people. The tragedy took place at a night club in Saint-Laurent-du-Pont, a village in a forested area about 20 miles (32km) outside Grenoble, France.

The high death toll in this disaster was entirely due to the lack of normal safety precautions and safety equipment in a dance hall that was little more than a barn in the middle of nowhere.

Club Cinq-Sept was a popular disco which drew young people with cars from as far afield as Annecy, Chambéry, and Grenoble. It had an interior gallery which ran around its perimeter and overlooked the dance floor and stage. The whole place was decorated in the psychedelic manner of the period – there were revolving colored light displays and its alcoves were decked out to look like grottoes. But it had no licence for dancing, it was badly lit, badly ventilated, had no mains water and no windows.

As a small fire turned into an inferno – fed by an array of cheap plastic furnishings – dense acrid fumes quickly filled the whole venue. The audience – who had been listening to a set by an up-and-coming group from Paris called The Storm – panicked and struggled towards the exits, but when they reached them they found that they had been padlocked to keep out gatecrashers. They banged desperately on the doors but could not break them down and there was no one to unlock them – the two managers who were keyholders were among the fire's first victims.

Most of the 30 people who escaped did so by squeezing their way back out through the "in" turnstile. Twelve of these survivors were badly burned. Nearly

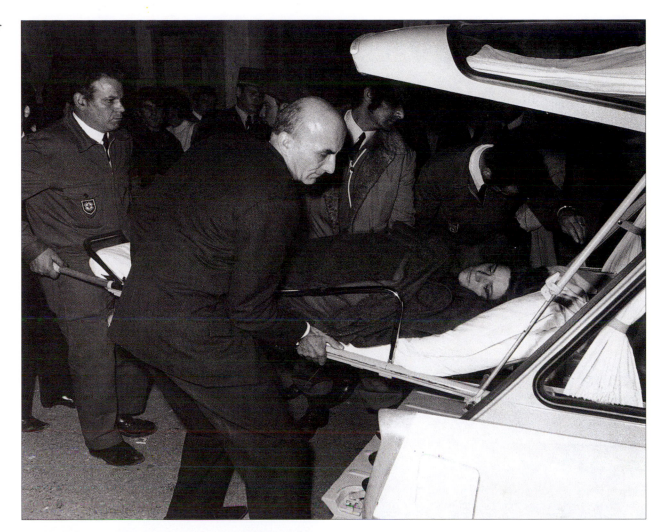

Right: A survivor from the dance-hall catastrophe is taken away to the hospital.

Below: Firemen searching the debris in the remains of the gutted Cinq-Sept night club.

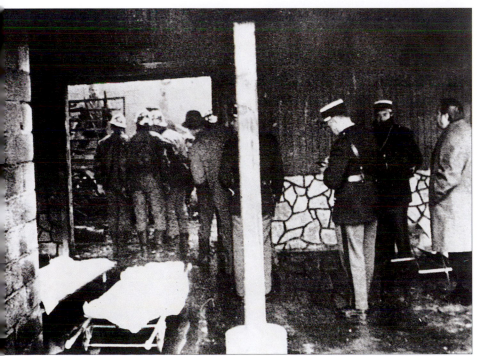

all the victims died through inhalation of toxic fumes and were asphyxiated before the flames reached them. Nevertheless, very few of the bodies were recognizable and most could be identified only by their jewelry, keys, and rings.

The one manager to survive the inferno was called Gilbert Bas. He afterwards described how he had been sitting in the club office when he had seen the alarm signal – a double flashing light – go off on his console. He at first assumed that it was the usual thing – another fist fight at the door – but when he got out onto the dance floor he heard cries of "fire!" He raced out of the club and drove straight to Saint-Laurent to raise the alarm. He felt there was nothing else he could do – Club Cinq-Sept did not even have a telephone.

Many fires gut the insides of a building but leave the exterior largely undamaged. The blaze at Club Cinq-Sept was different – by the time the emergency services arrived, even the corrugated iron roof had been melted by the heat of the flames. "The place went up like a matchbox," said one fireman sadly. "They didn't have a chance."

RADEVORMWALD, WEST GERMANY

MAY 27, 1971

Structural failures can also involve the possible misinterpretation of a signal, particularly if the nature of that signal is open to debate. This signal problem led to a head-on crash between a diesel railcar and a freight train. The normal procedure was for the driver of the diesel to take his train off the main line, wait for the on-coming train to pass, and then rejoin the main line to continue his journey. On this occasion it appears that the driver of the diesel commuter service was scheduled to wait for the freight-carrier at Dahlerau, and only proceed once given the all-clear.

He did wait at Dahlerau as scheduled, but then saw a sign from the stationmaster there that gave him permission to carry on, taking his passengers, a party of schoolchildren returning home after an educational outing to Bremen, to their destination. He entered the single line of track ahead and was hit by the second train. Casualties among the passengers on the diesel train were heavy. Forty-six were killed and 25 suffered varying degrees of injury.

The list of casualties might have been much longer had the stationmaster, who was aware that the freight train had not passed, not attempted to attract the commuter train driver's attention by waving a red warning light and, when this failed to have any effect, rang the emergency services before the collision had taken place. Rescue services were able to reach the crash site in quick time.

Below: Stretchers are brought up to the wreck to remove the dead and injured from the train's coaches. Many of the casualties were teenage schoolchildren.

FREIBURG-IM-BREISGAU, WEST GERMANY

JULY 21, 1971

Below: *Part of the "Schweis Express" lies at the bottom of the embankment where it came to rest after it had left the tracks as the train's driver attempted to negotiate a curved section of track at high speed.*

Although this incident, which left 22 dead, was in part the fault of the driver of the "Schweis Express," it also highlighted certain shortcomings in the way in which speed restriction signs were displayed throughout the rail network in the German Federal Republic. The accident occurred at night as the train was making its way from Basel in Switzerland to Copenhagen, the capital of Denmark, along the Rhine valley.

The driver attempted to negotiate a curve in the track at a speed close to 90mph (144km/h), more than twice the permitted rate, and the express left the track. Twelve carriages and the locomotive then crashed down a high embankment and smashed through an adjacent house. The two remaining coaches stayed on top of the embankment but slewed through 90 degrees to come to rest across both sets of tracks, blocking the stretch of line completely.

When the authorities investigated the Freiburg-im-Breisgau crash, they found it had remarkably similar characteristics to an event at Aitrang during the previous February. It soon became clear that the high speeds reached by the most modern trains coupled with night-time travel could lead drivers to miss the rather small signs showing that speed restrictions were in force. Changes were made to the safety system. Henceforth, the previous posts placed approximately every 1000 yards (923m) were superseded by larger trackside boards positioned no more than 250 yards (231m) apart. This procedure became standard throughout the West German rail network.

SEAWISE UNIVERSITY, HONG KONG

JANUARY 9, 1972

Despite its unfamiliar name, the *Seawise University* was in fact one of the most famous vessels of the 20th century. The largest passenger liner ever built, the *Queen Elizabeth* (as the ship was called for most of its life), was launched shortly before World War II. In fact, war broke out before the ship was completed. The *Queen Elizabeth* spent most of the war dodging German submarines and made a significant contribution to the Allied military effort. It transported over 800,000 soldiers between 1940 and 1946.

After the war the *Queen Elizabeth* was refitted to carry out its intended task – fast, luxurious travel across the Atlantic between England and New York. The *Queen Elizabeth*, along with the *Queen Mary*, which had also had a distinguished wartime career, enjoyed a brief moment in the limelight, but the truth of the matter was that the age of luxury sea travel was drawing to a close. Despite expensive refits and ever more opulent furnishings, the great liner was losing out to air travel. The *Queen Elizabeth* was sold to the United States but plans to turn it into a tourist attraction never reached fruition. In 1970 the liner, anchored at Florida's Port Everglades, was sold to C.Y. Tung.

C.Y. Tung had grand plans for the *Queen Elizabeth*. The ship was renamed and was to sail to Hong Kong where, at enormous cost, it would be transformed into a floating university. It was also envisaged that the ship would be used for cruises. Because of problems with the aging ship's boilers, the *Seawise University* took six months to sail from Port Everglades to Hong Kong, where it arrived in July 1971.

Neither of the new owner's schemes was ever achieved. While the extensive refit was being carried out in Hong Kong harbor, the ship was sabotaged. On January 9, 1972, it seems that several fires erupted simultaneously in different locations throughout the ship. The blaze was finally reported to the harbor authorities at 1030 hours by a helicopter flying over the

Left: The Queen Elizabeth *was the world's largest passenger liner, and the pride of its owner, the Cunard Line.*

Above: Fireboats battle with the flames as Seawise University *(the former liner* Queen Elizabeth) *burns in Hong Kong harbor.*

ship. The ship's own firefighting system proved inadequate to deal with the flames, which spread rapidly through five of the ship's 11 decks. The explosion of an oil tank then added to the conflagration. Those working on the vessel could do nothing more than flee for their lives.

When the Hong Kong firefighting services arrived to deal with the fire, they faced a virtually impossible task

– most of the *Seawise University*'s superstructure was ablaze and, as the hours passed, the liner began to list dangerously. The main centers of fire were finally extinguished at daybreak on January 10, but by this time the ship was too badly damaged to stay afloat. The once-great liner rolled over on its side, coming to rest in 40 feet (12m) of water. Dismantling of the vessel got underway in 1974.

Right: Despite the efforts of the firefighters, the liner keeled over, coming to rest on the seabed of the harbor.

THE KELLOGG MINE, IDAHO, USA

MAY 2, 1972

Right: Smoke pours out of the Sunshine silver mine at Kellogg, Idaho, after a fire started below ground in one of the worked-out seams.

All mining is hazardous, and the deeper the mine the greater the danger. The Sunshine Mine at Kellogg in the hills of Idaho, USA, is one of the deepest excavations in the world. It has been mined extensively for its silver, a precious native element that generally forms in long, thin veins. These deposits may extend for great distances underground, and over many years the excavation of the seams at the Kellogg mine has created a labyrinthine warren of diggings, some of which are so deep that miners can stay down there for no longer than 30 minutes at a time.

The fire that killed 91 men at the Sunshine Mine started in one of the many worked-out parts of the mine, 3700 feet (1100m) below ground. At depths like these, there is greatly increased pressure and temperatures typically in excess of 100°F (38°C). In such conditions, sparks may be created for no apparent reason. This phenomenon, known as spontaneous combustion, is thought to have happened at the Sunshine Mine at midday on May 2, setting fire to one of the supporting timbers. The fire spread quickly, but the smoke outran it into the ventilation shafts.

Throughout the rest of the day, a column of thick white smoke belched out of the mine's main exhaust stack at the pit head. Rescue teams wearing oxygen masks went down immediately to search for survivors in the smoky darkness, while engineers set about sealing off empty shafts to prevent the fire from spreading even further. Once the location of the fire had been identified, fresh air was pumped down into nearby parts of the mine to help any miners who might have taken refuge in neighboring galleries.

Many miners were brought out safely, but the bodies of 24 men were soon found lying at the bottom of the main elevator shaft. These victims included the elevator operators – one of the worst aspects of this tragedy was that no one else below ground knew how to operate the elevators and they could not be controlled

Right: On May 9, seven days after the outbreak of fire, an official from the Bureau of Mines prepares to ride a steel capsule down to the level where the fire started.

from ground level. It was thought that the remainder of the shift who were unaccounted for might have fled to safety in another part of the mine – but the mine contained hundreds of miles of tunnels and the question was, where were they? Further teams of rescuers were lowered down the main shaft in a capsule on a steel cable.

The search continued for a week, by the end of which 47 bodies had been recovered. Hope was fading, when rescuers found two men still alive and in good health 4800 feet (1440m) below ground. They had survived on the pumped-in air and by eating the packed lunches of the men who had died. This discovery raised expectations among relatives at the pit head, but shortly afterwards a massive cave-in severed lines carrying compressed air. No one else came out alive.

The subsequent inquiry slammed the safety measures at the Sunshine Mine, which had provided inadequate training and no protective chemical masks.

Below: A young miner, Tom Wilkenson, is brought out of the mine alive and well seven days after the fire started. He and another miner survived on the packed lunches of their dead workmates in an air pocket of pumped-in air.

THE SENNICHI BUILDING, OSAKA, JAPAN

MAY 13, 1972

Below: A busy street in the center of Osaka. The Sennichi department store in the city center caught fire late at night, killing 118 people and gutting the building.

A discarded cigarette or a small electrical fault was the most likely cause of a fire that gutted the seven-story Sennichi department store building in Osaka, Japan, killing 118 people in a rooftop night club. A number of people died when they jumped off the parapet to escape the flames, but most of the victims were found in the night club, having succumbed to smoke and poisonous gases. Some died so quickly they had not even time to panic – one body was found with a hand outstretched as if about to pay for a drink.

The fire broke out late at night on the third floor of the building where construction work had been in progress for some time. It spread quickly through the darkened upper floors and then burst out onto the top level in the Play Town Club, a busy hostess bar with cabaret floor show.

There was no warning – the fire suddenly surged through the floor and walls and took its first victims before they realized what was happening. Then there was panic, and terrified hostesses and customers began to stampede through the smoke and flames toward the main door. Many of them died before they could reach the exit – they were either overcome by smoke or trampled to death in the crush. Others frantically smashed windows and leaped to their deaths. Eyewitnesses watched helplessly as people clung to the window ledges for a few moments before crashing onto the street below. One of the first people to arrive

on the scene, a doctor, likened the carnage at the foot of the building to a plane crash.

Then the emergency services arrived in force, enabling 49 people to escape unhurt down fire brigade ladders. But the two tubular canvas chutes that were sent up from ground level to speed the evacuation proved disastrous – some people, too frightened of the smoke and flames to form a queue, tried to climb down the outside of the chutes but lost their grip and fell to their deaths; others who slid down the inside of the chute as they were meant to do, shot out at the bottom, which was unattended, and were killed by the impact.

The following day, an electrician who had been working on the third floor was arrested on a charge of negligence. The inquiry into the Sennichi fire found that although the building did have a smokeproof emergency staircase from top to bottom, it was unaccountably locked on the night of the tragedy.

Below: As flames and smoke swept through the Sennichi building, some people panicked and jumped to their deaths.

STAINES, ENGLAND

JUNE 18, 1972

The major reason for the crash of BEA Hawker Siddeley Trident on June 18, 1972, was the failure of the flight crew to make rational and effective decisions when a serious problem occurred during one of the most critical stages of any flight – take-off.

Below: Police and firemen survey the smoldering wreckage of the BEA Trident that crashed into a field in Staines shortly after take-off from Heathrow airport.

Although at 51 he was by no means an old man, the BEA pilot Stanley Key was by all indications not in prime physical condition. Six months prior to the crash he had undergone electrocardiogram screening and been passed fit to fly by medical examiners of the British Civilian Air Authority, yet the autopsy carried out on his body after the disaster showed that the arteries around his heart were severely restricted.

The Trident was due to make a scheduled flight from London's Heathrow airport to Brussels, Belgium. One element in the sequence of events that led to the tragedy can be traced back to the flight crew restroom at Heathrow, where Key became involved in a heated argument with a fellow pilot. The coroner concluded that the rise in blood pressure that this would have caused damaged one of the arteries of his heart.

By the time he boarded the aircraft he was probably in great pain. Both his first and second officers were in their early twenties, with far less experience than their captain, and were undoubtedly reluctant to question his command. Once the 112 passengers (almost the maximum capacity of the Trident) were seated and briefed, the flight crew proceeded as normal with their

Left: *The fuselage of the crashed plane reassembled at the Royal Aircraft Establishment, Farnborough, where it was taken for investigation.*

preflight preparations. The aircraft, given the call sign "Papa India," was cleared for take-off, and at approximately 1709 hours lifted off from Runway 28R at Heathrow.

Laboratory research has shown that all pilots experience a rise in heart rate at this point in a flight – this would have put further stress on Key's heart, probably causing him extreme discomfort, and this provides part of the explanation for the tragedy.

Only 75 seconds into the flight, one of the flight crew (most probably the captain) retracted the lift-assisting devices positioned on the leading edge of the wing. The Trident was, however, only traveling at 70 percent of the prescribed speed necessary for this action. The crew then reduced thrust, and retracted the lift-enhancing flaps on the trailing edge of the wing. With the Trident's insufficient airspeed the combination of these actions caused the aircraft to become dangerously unstable, but not critically so.

Although the aircraft was approaching a stall, the "stick-shaker" would have provided the crew with sufficient warning to lower the nose and redeploy the droops.

The Trident was one of the first aircraft to be fitted with the stick-shaker, which, as its name suggests, is a device that shakes the control column if the aircraft is approaching a stall. Later analysis showed that after it activated a second time, the device on Key's flight deck was manually overridden. Many Trident crews had expressed their dissatisfaction with the system, reporting that it was prone to engage unnecessarily. Someone on the flight deck of "Papa India" must have assumed incorrectly that the device was malfunctioning and switched it off.

Less than two minutes into the flight, the Trident entered an unrecoverable stall at a height of about 1970 feet (600m). It impacted at a relatively shallow angle in a field three miles (5km) southwest of the runway threshold, having just missed a major road packed with homebound commuters. There was no explosion, and the automatic extinguishing system rapidly doused a fire that erupted. There were no survivors among the 118 bodies pulled from the wreckage.

Unfortunately, there was no cockpit flight recorder installed on the Trident, which despite the crash remained largely intact. This added to the difficult task the air accident investigators faced when trying to explain the deaths of the 118 people on board.

Since that time, it has become almost routine for a cockpit flight recorder to be installed on the flight deck of a commercial airliner and the absence of it would now be considered negligence on the part of the airline. This is one more indication of how air safety improves after a disaster.

CHICAGO, ILLINOIS, USA

OCTOBER 30, 1972

A driver's failure to bring his "Highliner" double-deck electric locomotive to a pre-planned halt at a local station in Chicago, even after he had been given specific instructions to do so by his conductor before setting out on this journey on the Illinois Central Railroad, led to a violent collision between two trains that left 45 passengers dead and 330 injured, many seriously. The Chicago station in question, 27th Street, was known by railroad employees to be a regular "flag stop," meaning that commuter trains would halt there on request. However, stops were made on a more frequent basis, chiefly for the benefit of hospital staff who used the station during the rush hours.

On the late October day in question, the driver passed through the 27th Street station, ignoring his conductor's earlier request to stop the train to pick up passen-

Left: *Firemen cut through the twisted metal of the Illinois Central Railroad train involved in the rear-end collision at Chicago's 27th Street station.*

Left: The Michael Reese hospital where many of the injured were taken can be seen in the background of this photograph of the crash site. Many of its staff also used the service.

Below: Cutting equipment is deployed by the rescue services to slice through the twisted metal to gain entry into the carriages.

gers, but the "Highliner," a unit only recently introduced into service by the railroad, was then brought to a sudden halt some 200 yards (185m) farther down the track once the error had been recognized by the train's crew. At this point, the commuter train was also more than 125 yards (115m) beyond an automatic safety signal that would have warned the crew of any potential danger farther back down the track.

The staff of the "Highliner" then began slowly backing up their train with the intention of picking up any passengers that may have been waiting at 27th Street. This proved to be a serious error of judgment on their part. As the "Highliner" went into reverse for the return journey, it suffered a violent rear-end collision with the following train, which was scheduled to pass though 27th Street without stopping. This struck the "Highliner" at a speed of some 50mph (80km/h) and great damage was done to the rolling stock of both trains, particularly those closest to the point of impact.

The subsequent enquiry by the railroad also revealed that the driver of the second train was partly responsible for the closing speed at which the crash occurred. He had been given a yellow warning light at the previous signal which indicated that he was to reduce his speed to no more than 30mph (48km/h) before passing through 27th Street station.

This warning was not complied with, as the speed of the second train was nearly twice this when the collision happened, but the driver should still have been able to spot the reversing "Highliner" as it made its way back to the station to pick up passengers and bring his locomotive to a safe halt well before the crash took place. Nevertheless, the chief cause of the smash was the failure of the "Highliner's" crew to take adequate safety precautions before reversing their train into 27th Street station.

TENERIFE, CANARY ISLANDS, SPAIN

DECEMBER 3, 1972

Below: The island of Tenerife, which is a popular destination for Europeans on vacation. It was a charter flight full of German tourists that crashed fatally on take-off from the island's airport.

There was a huge increase in the volume of air traffic operating in and out of Tenerife during the early 1970s, due to the growing popularity of the island as a destination for Europeans on vacation, particularly from West Germany and the UK. During the winter months, when this accident occurred, the mild weather attracts countless numbers of sun seekers, and the volume of traffic remains correspondingly high.

The Spantax aircraft – a Coronado – was on a charter service scheduled to take off from Los Rodeos airport in the early dawn (an inconvenience that will be familiar to anyone who has ever taken a charter flight) bound for Munich in West Germany. Low overcast was covering the island, but weather conditions were not deemed serious enough to warrant suspending normal operations. In the previous six years Spantax had suffered two fatal accidents in the Canaries with the loss of 33 lives. In this tragedy more than 150 were to perish.

The Coronado took off at 0645 hours, climbed to a height of only 300 feet (100m), and then plunged to the ground. The fuselage came to rest in an inverted position. All 148 passengers, most of them German tourists, were killed together with the flight and cabin crew of seven Spaniards.

It was later concluded by the investigating Spanish authority that the pilot had lost control of the aircraft at about the point when the nosewheel left the ground. It proved difficult to pinpoint the exact reason why he experienced this loss of control, a problem that occurs in a considerable number of aircraft accidents that are attributed to human errors.

The most likely conclusion is that the pilot became disorientated in conditions of almost zero visibility. With the benefit of hindsight, his decision to take off when he was fully aware of the poor weather conditions might be thought questionable. That said, airline pilots are often under quite considerable pressures to maintain their schedules. All the time that commercial aircraft are sitting on the ground they are losing money for their operators, and this is clearly undesirable in a fiercely competitive business. These commercial pressures inevitably filter down to the captain, who is caught in the conflict between the profitability of his airline and the safety of his aircraft, passengers, and crew.

Above: The remains of the victims of the Coronado crash are loaded on to stretchers by Red Cross workers.

Right: The wreckage of the chartered Coronado airliner that crashed just after its dawn take-off, killing all on board.

SIBERIA, RUSSIA

MAY 18, 1973

Above: A Tupolev T-104 jetliner sitting on the tarmac. The hijack of a T-104 in the 1970s was the first occasion that a hijacking resulted in a crash killing all those on board – including the hijacker.

As with so many aspects of life in the former Soviet Union, the details of this disaster are shrouded in secrecy. The incident has the doubtful distinction of being the first occasion that a hijack resulted in a fatal crash. That it took place behind the Iron Curtain illustrates the fact that during the 1970s no country or people were safe from the air pirate.

The Tupolev T-104 was the first jet airliner to emerge from the Soviet Union and was introduced in September 1956. In May 1973 a Tupolev T-104A – one of Aeroflot's veteran fleet of T-104 jetliners – took off on a scheduled flight from Moscow to Chita, about 2920 miles (4700km) to the east. About 1500 miles (2400km) into the flight, the aircraft was at its cruising altitude of approximately 30,000 feet (10,000m) when the hijacker caught the attention of one of the cabin crew and announced that he was carrying an explosive device. He demanded that the jetliner reroute to China. His choice of destination, from one communist regime to another that was, if anything, even harsher, was curious. If he had landed in China, he would surely have been extradited to Russia.

Neither the hijacker nor his 80 victims reached their planned destinations. The Soviet newspaper *Pravda* reported next day that somewhere east of Lake Baikal the jetliner was fatally crippled when the hijacker decided to detonate the device. From that altitude, the chances of anyone surviving were infinitesimal.

SUMMERLAND LEISURE CENTER, ISLE OF MAN

AUGUST 2, 1973

Right: The Summerland indoor leisure center on the promenade at Douglas, Isle of Man, had been open barely a year before disaster struck.

Fifty people were killed when Summerland, a leisure center on the promenade at Douglas, Isle of Man, burned down in minutes after a small fire quickly became uncontrollable. It was the worst tragedy in Manx history.

Opened in 1972, Summerland was a state-of-the-art indoor tourist leisure center designed to provide "total entertainment" for families where parents could drink or dance on the top floor while their children played under supervision in other areas. It featured a wide range of attractions, including a basement funfair, discothèque, and solarium.

Summerland was an immediate hit with summer visitors and in its first year of operation managed to attract some 13 percent of the island's total tourist revenue. It was particularly crowded on the night of August 2, partly because it was a new and popular attraction, and partly because the day outside had been cold and drizzly.

It is estimated that about 2000 people were enjoying themselves in the complex when, just before 2000 hours on Thursday evening, a small fire – thought to have been caused by an electrical fault – broke out near a yellow, artificially thatched amusement kiosk on the first floor terrace. Eyewitnesses reported that within minutes the four-story building was ablaze from end to end. The fire brigade was called and reached the scene within four minutes, but they were already too late.

The complex had an overall roof made of acrylic sheeting. This US product, known as Oroglas, had been marketed as "slow-burning plastic," but it turned out to melt and burn very quickly indeed. At the height of the

Right: Within minutes of the fire starting, black smoke was pouring out of the fun complex as people inside sought to escape. Several of the emergency exits were found to be blocked or jammed.

Above: Acrylic sheeting lining the roof of the complex caught fire and burned quickly. Within 20 minutes the inside of the building was gutted.

fire, a huge pall of black smoke rose several hundred feet into the air above Douglas Bay and flames shot higher than the cliff face into which Summerland had been grafted.

Liquid plastic rained down onto the tourists as they made a terrible stampede for the exits. One said: "I saw a man with his hair on fire and the coat melted off his back and a youngster in his arms." Several emergency doors were blocked or had become jammed: some people were able to break the glass and climb to safety, but many were not so fortunate.

The whole center quickly became a fireball. Although every available resource on the island was deployed – 15 fire appliances and more than 100 firemen – the building was gutted within 20 minutes. In an emergency medical airlift, plasma and blood were flown from Liverpool on the English mainland, 80 miles (128km) away.

The subsequent inquiry into the causes of the Summerland blaze blamed the choice of construction materials, many of which would not have been permitted on the mainland. The Isle of Man, though part of the British Isles, is not a part of the United Kingdom – it has its own separate government and laws.

THE JOELMA BUILDING, SAO PAULO, BRAZIL
FEBRUARY 1, 1974

When an office building in the center of São Paulo, Brazil, mysteriously caught fire on the morning of February 1, 1974, many of the workers inside were trapped on the upper floors of the building beyond the reach of the fire brigade's longest extending ladders. Though this might suggest a very tall skyscraper like those in New York, the Joelma Building was only 25 storys high.

Left: As flames and smoke engulf the Joelma Building, a man leaps from the roof to his death.

Above: A helicopter lands on the roof to pick up survivors. The intense heat prevented helicopters from landing while the building was ablaze.

The bottom six floors of the Joelma Building were a multi-story parking lot, above which were 19 floors of offices, including the headquarters of the Crefisul Bank. Altogether, 600 people were working in the building that day, and there were also about 100 visitors and customers on the premises when – for no apparent reason – fire broke out on the 11th floor. Those closest to the exits – about 50 people – ran out with their clothes and hair on fire: they were more fortunate than many because, within seconds, every-one else inside the building had been cut off.

As the flames spread upwards, many people hurried upstairs. With the benefit of hindsight, this was a fatal error, but at the time it must have seemed the only rational thing to do. Those who had already been to the top of the building would have known that the highest levels were still unaffected by fire; others would have felt instinctively that it was too risky to try to get down through the fire to the ground floor. But the problem with this course of action was that the fire ladders could not reach more than halfway up the building.

At the time, São Paulo – a city of eight million people – had only 20 fire stations. Although the alarm was raised quickly there were insufficient fire tenders for the task and those that did answer the call were delayed by traffic jams and the crowds that had gathered in the surrounding streets to watch the blaze.

Meanwhile, the fire had taken hold of the Joelma Building, which was built of a cocktail of flammable materials. Many of the people still inside climbed out onto window ledges, desperately shouting for help. As the flames crept up on them, they faced a terrible choice between jumping and being burned to death.

When the rescue services finally arrived, they did all they could in the circumstances, and one brave fireman rescued 18 people whom he helped to safety along ropes which had been shot from nearby build-ings. Eventually, after four hours, the flames were brought under control, but by then 227 people had died and 250 others had been seriously burned or injured in other ways. Despite police investigations, the precise cause of the fire was never established.

PARIS, FRANCE

MARCH 3, 1974

Below: French rescue workers and police sort through wreckage of the DC-10 that crashed into a forest near Paris.

In the early 1970s the McDonnell Douglas DC-10 was rapidly acquiring a reputation for being an unlucky aircraft. This accident involving a Turkish Airlines DC-10 resulted in the death of all 346 on board. It was directly attributable to a design fault that had already caused two earlier incidents in which – miraculously – there had been no fatalities.

The rear cargo door of the DC-10 was designed to open outwards, and this meant that when the aircraft was at altitude and the cabin was pressurized, the rear door was being pushed outwards. In June 1972, the door of an American Airlines DC-10 had blown open, but the captain managed to land the aircraft safely. Following this accident, the US Federal Aviation Administration

prepared to issue an order that alterations should be made to the door locks. But for a variety of reasons, and even though the manufacturer was aware of the potentially catastrophic results if the door locks failed at altitude, the order was downgraded to a request that modifications be made.

Two years after the incident involving the American Airlines DC-10, a similar aircraft operated by Turkish Airlines took off at 1230 hours on Sunday, March 3, 1974, from Orly airport, serving Paris, bound for London Heathrow. The aircraft was filled almost to capacity, mostly with British tourists. As the aircraft climbed through 11,000 feet (3300m) over the village of Saint-Pathus, the rear cargo door failed. The aircraft suffered an explosive decompression, entered a steep dive, and impacted at nearly 500mph (800km/h) into a forest. There were no survivors.

Above: *Part of the fuselage of the Turkish Airlines DC-10 that crashed in the Forest of Ermenonville shortly after take-off from Orly airport.*

Right: *Rescuers carry away a body from the crashed aircraft. There were no survivors.*

FLIXBOROUGH, ENGLAND

JUNE 1, 1974

Above: The explosion at the Nypro works near Flixborough was the worst explosion in Britain since World War II. The region was declared a disaster area.

In the world's worst disaster in the chemical industry, 29 workers were killed and more than 40 injured when fire caused by an explosion gutted the £18 million ($29 million) Nypro nylon works at Flixborough, near Scunthorpe, Lincolnshire. Hundreds of acres of surrounding countryside were devastated by the resulting conflagration.

The Flixborough plant – which was owned jointly by the British National Coal Board (NCB) and Dutch State Mines – was at the time the main source of material for most British man-made textiles. It was used principally to make caprolactam, an essential component in the manufacture of nylon fibres. This material is obtained from cyclohexane, which in turn is a byproduct of coal-coking processes: it is highly flammable and has general properties similar to those of gasoline.

The seat of the explosion in the 20-acre (8 hectares) factory was an area known as Section 8 – it was here that the cyclohexane was oxidized by being heated and subjected to intense pressure. Nearly all the victims were working inside Section 8 or in the immediate vicinity – the dead were thought to have been killed instantly by a sudden surge of intense heat.

The disaster took place late on a Saturday night, when only 70 people were working in the entire plant. If the explosion had happened during a normal working weekday, the death toll would almost certainly have been much higher: Flixborough normally had 200 people on duty in every shift from Monday morning through to Friday evening.

Twenty-four hours after the explosion, a pall of smoke rising from the ruins of the factory was still visible more than 20 miles (32km) away. For days, emergency services could not actually extinguish the fire, which went on for almost a week before exhausting itself, rather than being extinguished. There was for a time a grave fear of radioactive fallout from a container of gamma-ray material that was known to be in the plant, but once the fire began to subside the canister was discovered intact in one of the less badly affected parts of the factory.

As the full scale of the disaster became clearer, it was discovered that the blast had been so great that it had rocked the foundations of houses in the village of Amcotte on the other side of the River Trent. Even at that distance from the explosion, serious structural damage had been caused to a number of substantial buildings: the pillars of the church were shaken, and the church was closed pending a full structural survey.

Water that had been used by the fire brigade to douse the flames lay in lagoons in the surrounding fields and was later found to be seriously contaminated with toxic chemical waste. Much of this liquid was eventually drained into the River Trent, only 100 yards (90m) from the plant: this led to serious pollution and threatened local wildlife.

Above: *A cloud of toxic smoke hides the remains of the chemical factory that exploded on June 1, 1974, killing 29 people.*

Right: *People from 100 houses near the factory were evacuated because their homes had been damaged by the blast.*

CARIBIA, GUAM

AUGUST 13, 1974

Dogged by misfortune throughout its life, the passenger liner *Caribia* had an ignominious end when it was smashed to pieces on its way to the scrap yard.

The *Caribia* was not the most fortunate of ships. It was launched as the *Caronia* in late October 1947 and began its working life on the Southampton to New York line. However, the growing competition from air travel was making things hard for the Atlantic passenger liners, and the *Caronia* was far from profitable. Although the ship was popular as a Caribbean cruiser, in 1967 the owners, Cunard, bowed to the inevitable and the *Caronia* was withdrawn from service.

The *Caronia* was sold in 1968 and was renamed the *Caribia*. After a refit in Piraeus its new owners decided to try their luck in the cruise market out of New York. But the change of name did not seem to improve the ship's fortunes. After the *Caribia* set out on its second cruise on February 28, 1969, a steam line in the boiler room exploded on March 5, killing a crew member. The ship lost power and was eventually towed back to New York and sold.

The *Caronia* had problems finding a permanent berth in New York. It finally fell foul of the harbor authorities and was fined for "parking" illegally! The difficulties of finding a berth were settled at the beginning of 1971 and the *Caribia* docked at Pier 56. The vessel was, however, a huge drain on the owner's financial resources and a way to dispose of it was sought. Because the *Caribia* was no longer considered a viable business, its owner finally accepted an offer from a Taiwanese scrap yard.

On January 25, 1974, the ship sailed from New York on its final journey. It was towed by the tug *Hamburg*. Bad weather forced the tug and its charge to put into Guam, but as the two entered harbor on August 13, the *Caribia* smashed into a breakwater. The ship capsized and broke into three sections. The remains were later broken up in situ.

Above: The Cunard liner Caribia *was built for the transatlantic route, but by the time the ship started sailing the route, air travel was making the passenger liners unnecessary.*

ZAGREB, YUGOSLAVIA

AUGUST 30, 1974

Above: Wreckage of the express from Belgrade to Dortmund lies scattered across the tracks at Zagreb station. Both its driver and his assistant received heavy prison terms for their part in the crash.

Two of those held responsible for this catastrophic derailment which left 153 passengers dead, the driver and his assistant, received prison sentences of 15 and eight years respectively. The length of their imprisonment reflected not only the long list of casualties but also the fact that they initially lied to the subsequent board of inquiry.

The disaster in the capital of the Croatian state of federal Yugoslavia involved an express making the long journey from Belgrade to Dortmund in West Germany. It was packed with native Yugoslavians returning to their places of work in West Germany after the summer holidays. The entire train was derailed when it attempted to take a curve on the approaches to Zagreb station at a speed of nearly 60mph (96km/h).

Regulations stated that the section should be negotiated at no more than 30mph (48km/h). It was also revealed that the driver and his assistant had ignored a stop signal shortly before the crash took place.

The locomotive's crew concocted a story that the express's brakes had failed at the crucial moment, but tests on them after the crash proved that they were in a perfect state of repair. The German authorities conducted experiments on rolling stock taken from a similar train to that involved in the Zagreb incident to also prove that the accident would not have taken place if the Yugoslavian crew had obeyed the speed restrictions. During their trial, the two Yugoslavians did tell the truth. They admitted falling asleep as they approached the curve outside Zagreb. Their tiredness and the derailment was caused by them working more than 300 hours in the previous month.

HOTEL FIRE, SEOUL, SOUTH KOREA

NOVEMBER 3, 1974

Eighty-eight people were killed and another 30 seriously injured in a fire that swept through a hotel in the eastern part of Seoul, the South Korean capital, during the early hours of a Sunday morning. The seven-story building had a night club on its sixth floor, and most of the dead were customers who were trapped inside it because the doors had been locked to prevent them leaving before they had paid their bills.

The blaze began when a hotel guest fell asleep while smoking in bed and the cigarette set fire to the mattress. By the time the fire was noticed it had taken a firm hold – it eventually raged for three hours before firemen brought it under control. By that time it had raced through the hotel and gutted the top two floors of the building.

Sixty-four bodies were recovered from the night club where more than 200 customers, mostly young people, had been dancing. Of the other victims, 13 were hotel guests who burned or were suffocated in their rooms, eight jumped to their deaths from upper windows, and three were found dead on the roof.

Seoul city authorities later removed the director of the regional office of public hygiene from his post on the grounds of negligence because the night club, which had been within his jurisdiction, had failed to close at 0200 hours as required by law.

Below: An aerial view of downtown Seoul, where the hotel was situated. The night club, where most of the dead were trapped, was on the sixth floor of the seven-story building.

MOORGATE, ENGLAND

FEBRUARY 28, 1975

Below: The sad scene at London's Moorgate station following the crash there in February 1975. Members of the public look on as rescue workers comb through the crushed wreckage underground for survivors.

This crash at Moorgate station in London left 43 dead, including the driver of the train involved, and more than 70 members of the public injured. The investigation that followed the incident concluded that the deceased driver was responsible for the crash, although the precise reason for the failure to stop was never identified. An autopsy did not reveal any medical problem that could have led to him failing to stop and X-rays of bones showed that his hands were on the train's controls at the time of the smash. Equally puzzling was the fact that no mechanical problems with the train were ever found.

In the mid-1970s, Moorgate was a busy station, serving both London's Inner Circle tracks as well as being the end station in the capital for a self-contained commuter line stretching out to Drayton Park, less than three miles (4.8km) to the north. There had been long-standing plans to develop the line. Its initial construction was such that it could take mainline rolling stock, so that travelers from the more distant suburbs could reach the City of London on the Great Northern Railway.

The initial expansion scheme was, in fact, never completed and a second development plan in the early years of the 1960s also came to nothing. Consequently,

Right: Police, station staff, and rescue workers look on as one of the survivors of the Moorgate train incident is transferred to a local hospital.

Below: The scene on the platform at Moorgate, showing the rear of the train. The front carriages were compressed to a fraction of their original length by the violence of the crash.

the rolling stock in use on the line at the time of the Moorgate incident was of considerable age.

On the morning of February 28, the driver in question brought his train, its six carriages packed with commuters, into platform nine at Moorgate. This end of the line was protected by an over-run at the end of the platform, which was some 70 feet (21m) long and ended with a 40-foot (12m) long sand-drag, designed to stop any overshooting train moving at relatively slow speeds before it collided with the end wall. However, the train was traveling at far too fast a pace for this safety measure to have little more than a minimal effect on the events which unfolded.

The train smashed into the end wall of the tunnel at such a high speed that the first pair of old carriages were crushed to less than half their normal length by the impact. The front half of the third carriage also suffered a similar amount of damage. It was in these front carriages that the majority of fatalities and injuries took place.

The mangled wreckage blocked the narrow and low underground tunnel, making the work of the rescue teams that rushed to the site all the more difficult. Such was the complex nature of the rescue that one passenger took more than 12 hours to be removed from the crash site, although he subsequently died from his injuries.

The accident at Moorgate station led to the introduction of a number of more modern safety devices on the network. Chief among these was a piece of equipment that came to be known as the "Moorgate Control." This is designed to stop a train automatically if a driver does not follow standard safety procedures for whatever reason.

WARNGAU/SCHAFTLACH, WEST GERMANY

JUNE 8, 1975

Left: The crumpled and twisted wreckage of the trains involved in the collision at Warngau brought about by errors made by three employees.

It is not only train drivers who can be responsible for accidents, as this incident demonstrates. At fault were station staff at either end of the section of line on which the collision took place. The severity of their misconduct can be gauged by the fines levied against the three people held at fault. Two stationmasters were ordered to pay 5000 Deutschmarks each and a clerk 2000. All three were also given suspended jail sentences for their pivotal role in a smash that left 41 dead and 122 injured.

Rescue workers were somewhat hampered in their efforts because the incident took place at dusk and the roads surrounding the crash site were blocked by locals returning to Munich after spending the weekend in the region's mountain resorts. Congestion was so bad that helicopters were used to shuttle the most seriously injured to hospitals in the city.

The single-track section of line in question lay south of the city of Munich in southern Germany and the two passenger trains involved were packed with people returning home after enjoying a day out. The staff at the two stations made errors which allowed the trains to enter the single track stretch at the same time.

However, the press were critical of the fact that, according to the publicly available timetable, the two trains were scheduled to arrive simultaneously at opposite ends of the section. However, as was pointed out, the working timetable available to railroad employees was somewhat different and the fact of the same arrival times in the public timetable was not a contributing factor in the collision.

NEW YORK, USA

JUNE 24, 1975

Dark skies gathered over John F. Kennedy airport in the late afternoon of July 24, 1975. A menacing thunderstorm brought pounding rain, violent and unpredictable winds, and precisely the kind of conditions in which the weather phenomenon known as a microburst can occur.

The crash of the Eastern Airlines Boeing 727 was one of three incidents that occurred in the United States within a decade of each other that were attributed to wind shear associated with microbursts. In a microburst a downward moving column of air can cause an aircraft to drop suddenly, sometimes with fatal results.

On the afternoon of July 24, Eastern Airlines Flight 66 was on a scheduled service from New Orleans, Louisiana. After they had contacted approach controllers at JFK, the flight crew were requested to land the aircraft on Runway 22. The pilots of two aircraft who had landed on this same runway only minutes previously had reported experiencing very strong wind shear. One reported nearly crashing. Despite this fact the tower controller did not consider it necessary to close the runway.

The Eastern Airlines Boeing 727 was about 1.2 miles (1.9km) from the end of the runway when it was suddenly lifted by an updraft of air. Without warning, this suddenly changed to a downdraft, and the supporting headwind dropped away. This caused the aircraft to descend rapidly, with the downdraft continuing to increase.

Still more than three quarters of a mile (1.2km) from the end of the runway, the 727 struck a row of tall approach-light gantries, one of which tore off the outer tip of the port wing. Plowing through more of the unyielding metal towers, the aircraft was virtually demolished before it even hit the ground, scattering flaming wreckage across a road barely 260 feet (80m) from the end of the runway. Of the people on board 109 passengers and six crew members perished; seven passengers and two stewardesses survived, all with serious injuries.

Left: Rescue workers sifting through the debris of the Eastern Airlines Boeing 727 crash on June 24, 1975, in which 115 people died.

SCHIEDAM, NETHERLANDS

MAY 4, 1976

Above: The lead carriage of the stationary commuter train which was involved in the Schiedam crash. Most of the casualties in the collision were found here.

The crash at Harmelen in the Netherlands on January 8, 1962, heralded the introduction of an automatic train protection system, but by the time of this collision, which left 24 passengers dead, only 25 percent of the Dutch railroad network had received the safety device. The incident at Schiedam in May 1976 hastened the adoption of the system and a decade later some 60 percent of the network had received the device.

The collision at Schiedam began when the international "Rhine Express" was delayed leaving the Hook of Holland, heading eastward on the righthand track. The train was switched to the lefthand track as it was nearing the Rotterdam–Den Haag line at Schiedam so that it could overtake a slower stopping train also traveling on the righthand track.

Most Dutch lines of the time had signals which permitted travel in either direction on a single track, so there should have been no difficulty in carrying out this simple switch. However, disaster struck when another passenger train heading westward on the lefthand track ignored stop signals and failed to halt before entering the section of line for the Hook of Holland. As the "Rhine Express" electric locomotive passed the now stationary stopping train on the righthand track, it was hit head-on by the second local passenger train also operating on the lefthand track.

Consequently, all three trains became involved in the fatal collision. The first passenger coach of the westbound train was crushed by the impact between the two trains heading in opposing directions on the adjacent track. It was here that the greatest number of casualties occurred.

PATRA, RED SEA

DECEMBER 25, 1976

The 3920-ton *Patra* began life as *Kronprins Frederik* in 1941, but because of World War II the ship did not make its first commercial voyage until 1946. The *Kronprins Frederik* was a fine, speedy vessel that worked routes between England and Denmark until it was sold to Arab Navigators in 1976 and renamed the *Patra*.

The new owners quickly put the *Patra* into service as a roll-on, roll-off ferry operating between Jedda in Saudi Arabia and Suez in Egypt. It could carry more than 350 passengers. The vessel's final voyage began at Jedda on December 25, 1976. On board were crowds of Muslim pilgrims returning to Egypt after having made the pilgrimage to the holy city of Mecca. Some 50 miles (80km) and five hours out from Jedda into the Red Sea it was reported to the *Patra*'s master, Captain Mohammed Shaaban, that there was a fire in the engine room and that the flames were spreading rapidly through the ship. Shaaban ordered the *Patra* to be abandoned and sent out an SOS.

Many passengers refused to get into the ship's lifeboats without their belongings, which they had been told to leave behind, until they were forced to do so by crew members carrying axes. A Soviet tanker, *Lenino*, along with several other vessels, headed towards the *Patra* and was able to save 201 passengers and crew. Another 100 perished and the vessel finally sank. Investigators concluded that the fire had been caused by a gas leak from an engine.

Below: The Danish ferry Kronprins Frederik, *which was later renamed the* Patra. *The ship was working as a roll-on, roll-off ferry when it was destroyed by fire in 1976.*

GRANVILLE, SYDNEY, AUSTRALIA

JANUARY 18, 1977

Above: An injured passenger is gently winched from one of the carriages damaged by the derailment and bridge collapse.

Australia's rail network has not been immune from rail disasters. There were several incidents in the early days of the network that led to a significant loss of life. Among these were a rear-end collision possibly brought about by brake failure or signalling errors at Braybrook Junction on April 20, 1908, which cost the lives of 44 and injured close to 150, and an incident between runaway rolling stock and an express at Murrulla in New South Wales on September 13, 1926, which left 27 dead and 46 injured.

However, this incident in Sydney, the vibrant capital of New South Wales, was a much more severe affair and caused a good deal of disquiet among the city's inhabitants. Some members of the public were so keen to find the guilty party that they sent the driver totally unjustifiable death threats. As events were to prove, he

played no part in the cause of the tragedy. A simple but costly mechanical failure was the reason behind one of Australia's worst accidents.

The incident involved a commuter train that was heading for Granville station located on the outskirts of the city. The train was traveling at 20mph (32km/h), a wholly acceptable rate on the section of track where the crash took place. However, a derailment occurred and several of the carriages left the track as the train was passing through a bridge that carried local traffic.

The violence of the derailment was sufficient to send two of the effected coaches crashing into the piers supporting the road bridge with such force that it collapsed on to the train. Steel and concrete smashed down on the carriages, crushing them, and causing the vast majority of the casualties. Several cars crossing the bridge at the moment of collapse also fell into the chasm, adding to the horror which unfolded.

Rescue workers faced a dangerous and complex task to search through the wreckage for survivors and bring out the many dead. They had to proceed cautiously to avoid any further collapse and were not able to complete their operation until the following evening. Heavy lifting equipment had to be ferried to the scene to help in clearing the wreckage.

Unfortunately, the rescue efforts were also hampered by the thousands of onlookers who had turned up to witness the disaster and who had to be controlled by hundreds of police so that the rescuers could continue their delicate work. By the end of the rescue, the emergency agencies could confirm that 83 commuters had been killed and over 200 injured.

Initial suspicion as to the cause of the crash focused on the driver. He, however, was found to be totally without blame by a board of inquiry. Its members identified the poor quality of the track near the bridge as being the cause of the derailment.

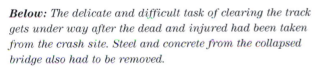

Below: The delicate and difficult task of clearing the track gets under way after the dead and injured had been taken from the crash site. Steel and concrete from the collapsed bridge also had to be removed.

Above: Several cranes have been positioned above the crash scene, ready to complete the task of removing the wreckage.

TENERIFE, CANARY ISLANDS

MARCH 27, 1977

Above: Of the two jumbo jets involved in this horrific collision on the runway, one was a KLM Boeing 747 like the one shown in this photograph.

As word of this accident spread across the holiday island of Tenerife in the early evening of Sunday, March 27, 1977, local hospitals were inundated with people generously offering their assistance as blood donors. Despite their efforts, and those of the emergency crews who rushed to the scene of carnage at Los Rodeos airport, this infamous disaster remains the world's worst ever commercial air accident. On the runway the inferno caused by the collision of two giant aircraft was still burning two days later. In all 583 people perished, on an island that was no stranger to air tragedy.

Both aircraft, the Royal Dutch Airlines (KLM) Boeing 747 (named *Rhine River*) and the Pan American 747 (named *Clipper Victor*), had been rerouted to Los Rodeos, Tenerife, after a bomb attack at Las Palmas, on the neighboring island of Gran Canaria, their scheduled destination.

The Dutch aircraft was on a service from Amsterdam, with a complement of 246 passengers and 11 crew members. Commanding the flight was one of KLM's most senior captains, Jacob van Zanten, assisted by First Officer Klass Meurs and Flight Engineer William Schreuder. The Pan Am aircraft had originated at Los Angeles, California. During a stopover at New York the crew of nine was changed and a further 103 passengers boarded for the flight to Las Palmas, bringing the total number of passengers to 378. Both aircraft were approaching the end of their flights when the airport at Las Palmas was closed due to the bomb attack. Both crews were told to divert to Tenerife, about 70 miles to the northwest. They landed within 30 minutes of each other early on Sunday afternoon.

It became clear that the airport at Las Palmas was not going to open in the near future, so the Dutch captain disembarked his passengers. He was also aware that the delay could mean his crew would exceed the time limit on their permissible duty hours, and he contacted the KLM operations center at Amsterdam to voice his concerns. The Amsterdam center confirmed that if the crew did not leave Las Palmas for the return flight by 1900 hours that evening they would be in breach of their duty hours limit. Since to do this knowingly would lead to a criminal prosecution, this is something that few experienced pilots would ever risk.

The Pan Am *Clipper Victor* touched down at Los Rodeos at 1415 hours but the captain, Victor Grubbs, elected to keep his passengers on board. Because of the increased volume of traffic at Los Rodeos that day, and the normally heavy weekend schedule, aircraft were parked in any available space. Both the empty *Rhine River* and the *Clipper Victor* were parked just off Runway 12 in a designated holding area (an area where taxiing aircraft wait until the crew are cleared for take-off).

Only 15 minutes after the *Clipper Victor* landed, Las Palmas was reopened, and aircraft at Los Rodeos began to leave for the short flight to the neighboring island. Frustratingly for the crew of *Clipper Victor*, they had been parked behind the Dutch aircraft. Until the passengers had been reembarked the American plane could not take off. Even when this lengthy embarkation process had been completed, news came through of delays at Las Palmas.

The Dutch captain then decided to refuel *Rhine River* at Los Rodeos, in an effort to speed up the turnaround at Las Palmas and stay within his tight schedule. This annoyed the captain of *Clipper Victor*, who had been on duty for over 10 hours by the time this process was completed. As the afternoon wore on the weather at Los Rodeos began to deteriorate, with light rain and fog reducing visibility at times down to 330 yards (300m).

Below: The Pan Am 747 in flames after it collided with the KLM 747 while taxiing along the runway in foggy weather.

It was almost 1700 hours before both aircraft were ready to depart. Because of the wind direction the crews were requested to taxi from the holding area to the far end of the two-mile-long (3.2-km) runway. Visibility had deteriorated to a point where it was difficult for the Dutch crew taxiing down the runway to ascertain their position, and for the Spanish tower controller to even see the aircraft. *Rhine River* reached the end of the runway, turned 180 degrees, and radioed to the tower that it was ready for take-off.

The next series of events were undeniably the cause of the tragedy, and can in part be attributed to fatigue, but in the main to human error. Without receiving clearance for take-off, or perhaps because they mistakenly thought they been given it, the flight crew of *Rhine River* began their take-off run.

On the flight deck of *Clipper Victor*, which was still taxiing in the opposite direction down the runway, the crew were anxious to get clear of the runway. The horrified crew then saw the giant Boeing hurtling out of the gloom toward them. Grubbs threw open the throttles and desperately tried to clear out of its path, as the first officer screamed "Get off, get off, get off."

Right:
Investigators
sifting through
the wreckage after
the worst ever
accident in
commercial
aviation history.

Below: The
charred fuselage
engine and wheel
sections of the
crashed KLM 747.

On the flight deck of the *Rhine River*, the Dutch captain yanked back on the control column, and grounded the tail of his aircraft heavily as he tried to pull up. Despite the actions of both crews, the undercarriage and the outboard engine on the port wing of the Dutch *Rhine River* hit the upper fuselage of the *Clipper Victor*, ripping most of it off. The *Rhine River* flew on for about 500 feet (150m) before sinking back onto the runway. Skidding down the tarmac for another 330 yards (300m), it came to a halt and was immediately engulfed in a huge fireball. Fire crews could do nothing to save any of those on board.

Of those on board the Pan Am aircraft, 70 people were pulled from the wreckage, but nine of them succumbed to their injuries in the following weeks, bringing the total terrible death toll to 583.

LEBUS, EAST GERMANY

JUNE 27, 1977

Before the reunification of East and West Germany in the late 1980s, information on train collisions or other rail accidents in the communist German Democratic Republic was difficult to find because of the state's strict control of the media. However, this incident was somewhat unusual in that Western journalists were allowed to visit the site of the crash later during the same day. Their reports indicated that the smash was the product of faulty routing.

Early on the morning of June 27, a steam-locomotive express was heading to the port of Stralsund on the Baltic coast from the town of Zittau on the border with Czechoslovakia, a distance of approximately 300 miles (480km). The collision took place close to the border with Poland, and the northbound express hit a diesel freight train while traveling at speed.

The violence of the high-speed impact initiated a series of ferocious fires that totally gutted the two locomotives involved and blocked the tracks in the vicinity of the crash with burning wreckage for several hours. The dangerous conflagration was finally brought under control by the East German emergency services. Nevertheless, the remains of the two badly damaged trains were still smoldering when the contingent of foreign journalists arrived on the scene to cover the story.

Subsequent investigations by the East German authorities revealed that the Stralsund-bound express had been incorrectly routed on to the wrong line as it was about to pass through the junction at Bossen, thereby ensuring that it would collide with the on-coming freight train. This seemingly simple error in switching procedure led to the death of 29 people on the two trains.

Below: The wreck of the passenger train involved in the collision near Lebus station gives some indication of the ferocity of the impact and the subsequent fire.

JOHOR BAHARU, MALAYSIA
DECEMBER 4, 1977

The perpetrator of this crime was clearly mentally unbalanced. There can be no other explanation for his actions, which resulted in his death and, more importantly, the death of 100 innocent people. The Boeing 737 that was destroyed in this incident was one of the fleet operated by Malaysia-Singapore Airlines on internal routes. It took off from Pinang in the north west of Malaysia on a scheduled flight to the capital, Kuala Lumpur, some 190 miles (305km) to the south.

The flight proceeded without incident until the flight crew were making their approach to Kuala Lumpur. At this point a man brandishing a revolver forced his way onto the flight deck and, threatening the pilot, demanded that the aircraft proceed to Johor Baharu,

the airport serving Singapore. The crew followed his instructions to the letter. Even so, as the Boeing approached Johor Baharu the hijacker shot and killed both the captain and first officer, the only two people on board capable of flying the aircraft. Pitching out of control into a near vertical dive, the aircraft impacted in a swamp about 30 miles (50km) southwest of Johor Baharu, exploded and broke up. There were no survivors among the 93 passengers and seven crew.

The incident threw a harsh spotlight onto security procedures at Pinang, and led to many questions being asked as to how the hijacker had been able to smuggle a weapon onto the aircraft. It has often been the case that security on internal flights is much laxer than on international routes. On this occasion it proved to be tragically inadequate.

Below: A Boeing 737 belonging to Malaysia-Singapore Airlines – similar to the aircraft that was hijacked and crashed on December 4, 1977.

AMOCO CADIZ, ENGLISH CHANNEL

MARCH 16, 1978

The *Torrey Canyon* oil spillage of March 1976 was still fresh in many people's minds when, less than two years later, disaster struck the supertanker *Amoco Cadiz* as it was making its way fully laden from the Persian Gulf to Rotterdam. The 288,513-ton supertanker sailed under the Liberian flag and was carrying around 250,000 tons of heavy crude oil. The tanker never completed its journey.

The *Amoco Cadiz* negotiated the Cape of Good Hope (it was far too large to be able to sail through the Suez Canal) and then sailed up the coast of Africa without encountering any problems. As the tanker headed up the west coast of France, it approached a congested area of sea some 30 miles (48km) off the northern tip of Brittany. Beyond lay the busy waters of the English Channel. Regulations were in place here so that both north- and southbound shipping had to stick to their own channels. This so-called "traffic separation" scheme was intended to prevent collisions.

The *Amoco Cadiz* entered the zone at a time of year when gales are common and can be ferocious. A supertanker under power should have been able to cope with the conditions, however. On the morning of March 16 the *Amoco Cadiz* had to maneuver to enter the northbound channel of the separation scheme. Shortly before 1000 hours the vessel's steering gear suffered a total failure – the rudder jammed hard to port. The tanker's master, Captain Pasquali Bardari, reacted quickly. He stopped the tanker's engines, hoisted a "not under control" signal, and sent out a radio signal to warn other vessels of the danger.

A German salvage tug, the *Pacific*, heard Bardari's distress call and hurried to offer help. However, Bardari had a problem – he needed to contact a senior official from the ship's owners in Chicago for permission to accept a tow. It was the middle of the night in

Below: The supertanker Amoco Cadiz wrecked on the rocks off the Brittany coast.

Right: The
Amoco Cadiz
*broke in two a few
days after going
aground. Here the
wreck is seen
awash in a sea of
its own black oil.*

Chicago and permission was not given until 1545 hours. The first tow rope from the *Pacific* broke and a second, attached to the stern, failed to halt the A*moco Cadiz*'s drift. Even its own anchors failed to stop the ship. At 2100 hours the tanker ran aground off the coast of Brittany. It was high tide and as the sea ebbed the vessel settled on to the rocks with its hull holed.

Crude oil poured out of the *Amoco Cadiz* for months. Despite the efforts of various agencies, deter-gents and booms failed to staunch the flow of oil. By March 19 the slick stretched for nearly 20 miles (32km). The French resorted to bombing the wreck, which by this time had broken in two. But by the time the *Amoco Cadiz* was swept away in a storm in March 1979, it had leaked over 250,000 tons of crude oil. Rough seas did, in fact, contribute to the dispersal of the oil. But not before the coastline of Brittany and its wildlife had been devastated.

*Right: Cleaning-
up operations
begin on the beach
of the little port of
Portsall, Brittany.
It was to be many
months before
the coastline
of Brittany
recovered from
the disaster.*

SAN CARLOS DE LA RAPITA, SPAIN

JULY 11, 1978

Above: The scene at the Los Alfraques camp site in Spain after a tanker carrying liquid propylene crashed on to the site and exploded, devastating the area and killing almost 200 campers.

Nearly 200 vacationers – most of them French or German and many of them children – were killed when a tanker truck with a cargo of liquid gas caught fire and then exploded after crashing into a crowded camp site on the Mediterranean coast of Spain.

It was about 1515 hours on a hot summer's afternoon when a truck hit a cement wall on the coast road at San Carlos de la Rapita, about 100 miles (160km) north of Valencia and 120 miles (192km) south of Barcelona. The driver lost control and his vehicle careered off the coast road and plunged down a hillside into the Los Alfraques camp site.

The truck, owned by the Cisternas Reunidas company, was believed to have been traveling along this slow, twisting route only in order to avoid paying the 1000 peseta toll on the turnpike which had recently been built to bypass this whole tourist seaside area.

Its tank was full almost to capacity with propylene – a gas used in the manufacture of alcohol and transported in the form of a pressurized liquid. Although it is highly inflammable, propylene is supposed not to be explosive. This puzzled crash investigators until eyewitnesses reported that they had seen flames coming from the truck before it went out of control – the propylene had evidently been heated by the fire on board for some time before it exploded.

The force of the blast was so great that it left a crater 20 yards (18m) across and hurled some of the campers 100 yards (90m) into the sea. It completely destroyed tents, a discothèque, and 12 holiday chalets. It also set off a chain of minor explosions in the bottled gas used by campers and the gasoline tanks of cars.

Right: A burned-out caravan on the camp site. The explosion completely destroyed tents and caravans, along with 12 chalets and a discothèque.

The death toll was high because at the time of the crash most of the campers were cleaning up after lunch or taking their siesta. If the accident had happened earlier in the day the death toll would almost certainly have been lower, because most people at the camp site would still have been swimming or sunbathing on the nearby beach.

Volunteers at the scene of the disaster said that identification of the bodies was virtually impossible because the explosion had dismembered them and the flames charred them beyond recognition. Nearly everyone was wearing swimsuits, which concealed little and protected even less flesh, so that most of the injured had extensive burns to between 50 and 90 percent of their bodies. The response to the disaster was immediate. The Spanish government put Air Force aircraft and helicopters at the disposal of the local authorities to evacuate the most serious cases. Cars and buses were commandeered to take the injured to clinics and hospitals all along the coast. Radio stations broadcast emergency appeals for blood donors.

So many of the injured were from Germany that the German government made arrangements the same day to airlift doctors and medical equipment to Spain from Stuttgart. Other nationalities among the victims included Belgian, British, and Dutch.

Right: Many of the campers were taking their afternoon siesta when the truck careered into their midst and exploded. Some were killed outright while others were hurled by the blast into the sea.

ABADAN, IRAN

AUGUST 19, 1978

Of all the violent incidents in Iran during the year that preceded the Islamic Revolution and the overthrow of the Shah in January 1979, none aroused greater outrage than the firebombing of the Rex Cinema in the southern oil city of Abadan. A total of 377 people in the audience were killed when terrorists soaked the floors and walls of the cinema entrance with petrol and then set fire to them during a late-night film.

The fire was started by Shi'ite extremists and was the worst in a long series of attacks on Western-style businesses such as banks, liquor stores, and restaurants, which were regarded by revolutionaries as a pernicious influence on the native culture. Indeed, this was the seventh time in a week that an Iranian cinema had been firebombed – in one of the earlier incidents,

in the holy city of Mashad, east of the capital Tehran, three members of the audience had been killed. The wave of atrocities took place during the holy month of Ramadan, when Muslims are supposed to fast and pray. The purpose of the outrages was to raise political tensions in the run-up to the anniversary of the assassination of Imam Ali, who is second only to the Prophet Muhammad as an object of veneration in Iran.

After the fire, the caretaker of the Rex was arrested for negligence. The police maintained that he was drunk on duty, but the real reason for the high death toll was that the doors were all locked shortly after the film began, probably in order to prevent latecomers getting in without paying.

In the circumstances, it is amazing that anyone came out alive, but 100 cinemagoers somehow managed to break their way out and escape from the

Below: The outside of the Rex Cinema in Abadan after the firebombing.

Right: The auditorium of the cinema after the fire. Within minutes of the torching, the cinema became an inferno in which 377 members of the audience died.

Below: Distraught relatives of the victims mourn their dead.

inferno uninjured. Witnesses later testified that the arsonists had taken time and trouble over their work and had left the petrol to seep in to the carpet before setting it ablaze. Members of the audience said that they had heard loud explosions just before flames shot up the cinema walls. At this point there was an immediate panic, and many people were trampled to death underfoot in the rush to escape. Others were burned to death or asphyxiated by toxic smoke fumes which belched out of the seats and other plastic furnishings. Nearly all the bodies were found in heaps behind doors which had not given way despite the frrenzied battering they had taken.

Police later said that most of the victims were so badly burned that it was impossible to identify them, and they were all buried without due ceremony in a communal grave outside the city. This form of burial was against strict Islamic laws, and the police station was later besieged by 2000 angry and distraught bereaved relatives demanding decent treatment for the bodies of their loved ones.

O'Hare Airport, Chicago, USA

May 25, 1979

Until the Lockerbie disaster in December 1988, this crash of an American Airlines McDonnell Douglas DC-10 was the worst accident involving a US carrier. The incident did nothing to enhance the reputation of the DC-10, and led to the suspension of its certificate of airworthiness by the Federal Aviation Authority.

The story of the disaster reads like a catalog of errors, beginning in the McDonnell Douglas plant, carrying on through the service life of the aircraft, and ending in a fireball one May afternoon with the loss of 273 lives.

The American Airlines DC-10, designated Flight 191, was due to fly a scheduled service between Chicago O'Hare and Los Angeles International airport. After the embarkation of the passengers and preflight checks were completed, the aircraft taxied to Runway 32 and at 1502 hours began rolling. At about 160mph (257km/h) the officer who was operating the radio

called "vee-one" – denoting the point at which the DC-10 was committed to take-off. Two seconds before lift-off, the aircraft shuddered violently, and the port engine seemed to lose power.

Observers on the ground saw smoke pouring from the left wing. Unknown to the flight crew, the port engine had detached completely and ripped a large section of the leading edge away, together with vital hydraulic and electrical lines. The flight crew – unaware that the engine had fallen off – carried out standard procedure for engine failure. The first officer pulled back on the control column, lifted off the runway, and proceeded to climb.

Because of the failure of many of the generator systems, the crew still had no indication of the seriousness of their predicament. About 20 seconds after lift-off, at an altitude of less than 300 feet (100m), the left wing suddenly stalled (due to the fact that the lifting slats had retracted because of the loss of hydraulic

Right: Debris from the American Airlines DC-10 scattered over the trailer park where it plunged into the ground.

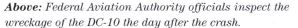
Above: *Federal Aviation Authority officials inspect the wreckage of the DC-10 the day after the crash.*

Above: *Firemen probe through the debris of the crashed DC-10.*

power). As the nose dropped below the horizon, the first officer sought in vain to wrest the jetliner from its dive toward the ground.

Thirty-one seconds after lift-off, the aircraft plunged into the ground near a trailer park, instantaneously killing all those on board, as well as two people on the

ground. Examination of the wreckage, and of the engine that had been thrown onto the side of the runway, revealed severe cracking.

All DC-10 aircraft in operation were grounded and subjected to careful inspection. Ten of them also revealed fatigue cracking in the wing engine mounts. The damage had been caused by hasty servicing procedures carried out at a number of airlines when the engines were removed from the aircraft for overhaul. At that time, airlines were constantly searching for ways to reduce servicing man hours.

Serious though the loss of an engine was, such an incident had occurred before with the successful recovery of the aircraft involved. However, the main contributing factors to the crash were the retraction of the leading edge slats, which provide essential lift during the early stages of the flight, and the fact that the flight crew decided to reduce power to achieve optimum climb speed. If they had maintained speed, and if the slats had locked mechanically, the accident might well have been averted.

ATLANTIC EMPRESS, CARIBBEAN

JULY 19, 1979

The loss of the 292,666-ton *Atlantic Empress* supertanker was the result of a collision with another monster supertanker, the 210,257-ton *Aegean Captain*. Both tankers were Liberian registered. The *Aegean Captain* had recently filled up with crude oil in Venezuela and was bound for Singapore, while the *Atlantic Empress* was making its way from Beaumont, Texas, to the Persian Gulf. The combined loss of crude oil into the Caribbean was estimated at about 280,000 tons. Lloyd's of London, the maritime insurers, described the disaster as "our biggest marine loss."

The collision occurred a little way off the island of Little Tobago in the early evening at a time when the weather was worsening. As night fell visibility in the area began to drop quickly, yet although both ships had modern radar systems, no one on watch on either ship appears to have spotted that the two supertankers were closing fast. It was not until the two ships were about one mile apart that lookouts spotted the danger. By then, however, it was far too late to do anything. Supertankers take a long time to stop or maneuver and neither option was available at such a short distance. The tankers struck each other with a combined speed of 30 knots.

It was just after 1900 hours when the *Aegean Captain*'s bow crashed into the *Atlantic Empress*'s port side. Fires were ignited on both vessels. The crew of the *Aegean Captain* were able to make an orderly evacuation of their vessel, but things did not go so smoothly on the *Atlantic Empress*. The crew had difficulty launching their lifeboats and of the 42 people on board the tanker 26 died. The *Aegean Captain* was towed back to Curacao in the Dutch Antilles, but the *Atlantic Empress* sank.

Below: Clouds of dense smoke billow from the doomed Atlantic Empress *after its collision with another supertanker,* Aegean Captain.

ROSS ISLAND, ANTARCTICA
NOVEMBER 28, 1979

The report of the commission set up to investigate this tragedy, which was to date New Zealand's worst ever air disaster, made scathing attacks on both the carrier and the flight crew. Both parties certainly made disastrous errors of judgment that turned what had been planned as a pleasant sight-seeing trip over the stunning landscape of the Ross Ice Shelf in Antarctica into a catastrophe.

Flying in the Antarctic requires constant vigilance, highly developed skill, and precise judgment, both to predict and safely assess the often hostile weather conditions, and also to navigate successfully without the benefit of obvious ground reference points. Another complication is that magnetic compasses are rendered useless in Antarctica because of the existence of powerful magnetic fields around the South Pole.

Left: Antarctica is stunningly beautful when seen from the safety of the air, but in reality it is a desolate, inhospitable wasteland of ice and frozen seas. It was in a landscape like this that the McDonnell Douglas DC-10 crashed in November 1979.

The flight officers of the Air New Zealand McDonnell Douglas DC-10 were given a briefing prior to the flight, which was later judged to have been insufficiently detailed. Only one of the five officers had made such a flight before. During preflight preparations at Auckland, one of the crew inputted a detailed flight plan into the inertial navigation system on the DC-10. This plots a flight profile and is used to calculate the position of the aircraft and where possible cross-references with land-based navigation aids.

Investigations later revealed that this data was wrong. It was based on incorrect information that Air New Zealand aircraft had been using for at least 14 months. However, as all previous sight-seeing trips had been made in conditions of far greater visibility than on the day of the fatal crash, the flight crews would not have been forced to rely heavily on their instruments, as was the situation on this day.

As he approached Ross Island, which lies at the edge of the Antarctic ice cap, southwest of New Zealand's South Island, the pilot decided to descend below the base of the low, dense overcast, to facilitate better viewing for the passengers.

This was a reckless violation of the guidelines, which strictly forbade flying below 6000 feet (1800m) until the mountain range that was now lying directly in the path of the airliner had been passed.

The error in the inertial navigation system data also meant that the aircraft was some 30 miles (50km) off course and on a track headed toward the towering peak of Mount Erebus.

Flying in thick cloud the flight crew could see nothing of the approaching danger, and decided to descend to 1500 feet (500m). As the aircraft emerged from the dense overcast, visibility was still poor, and the crew saw the wall of rock and ice in front of them too late. The aircraft struck a slope on the western approaches to Mount Erebus at nearly 300 mph (483 km/h), killing all 257 persons on board.

Left: The wreckage of the Air New Zealand McDonnell Douglas DC-10 strewn on the snow-covered slopes of Mount Erebus in Antarctica. No one on board survived.

Between 1980 and 1989 the world suffered some of its worst disasters in living memory, on land, at sea, and in the air. Air catastrophes were particularly noteworthy. Among the worst was the total destruction by a terrorist bomb of Pan American Flight 103 over Lockerbie, Scotland, in late December 1988, which left some 260 people dead. At sea, there were equally disturbing losses, not least the sinking of the ferry *Herald of Free Enterprise* in Zeebrugge harbor, Belgium, in March 1987. Because of human error, some 200 people died in the incident. Equally severe events occurred on rail systems, and fire again proved to be a deadly enemy.

The period was littered with many similar events, and it became increasingly clear that the wider public was losing its faith in those appointed to deal with matters of safety. Many ordinary people and pressure groups began to feel that these key individuals, those who owned or ran companies, were putting the public in danger in their headlong rush to secure ever greater profits for their companies, leaving the often overstretched and underfunded rescue services to pick up the pieces in the aftermath of a disaster. Others argued that the finances that should have been earmarked for technological upgrades were being used to pay dividends to shareholders rather than being spent on improving services. Despite attempts to calm these genuine public fears, businesses found it difficult to shake off the accusations.

Background: In an attempt to reduce the pollution caused by oil spillage from the Exxon Valdez *in 1989, a second tanker is run alongside the supertanker to empty its storage tanks.*

1980-1989

RIYADH, SAUDI ARABIA

FEBRUARY 19, 1980

Above: The Saudi Arabian airliner in which 301 people burned to death was a Lockheed Tristar like this one.

When the fire alarms indicating smoke in the aft cargo compartment sounded on February 19, 1980, on the flight deck of a Saudi Lockheed Tristar bound for Jiddah in Saudi Arabia, it quickly became obvious that the flight crew had little idea what to do.

The Tristar was only minutes into the second leg of a service from Karachi, Pakistan. The crew, none of whom had much experience on this type of aircraft, delayed four long minutes before confirming the alert, and then – almost unbelievably – had to search for the manual showing the correct procedure in such an incident. Clearly failing to treat the fire with adequate seriousness at this stage, the flight engineer repeatedly suggested that there was "no problem."

In the passenger compartment, the 287 passengers and 10 cabin staff were somewhat more concerned. As smoke seeped into the cabin, the cabin crew struggled to maintain calm. The captain turned back to Riyadh to land. After landing, the captain should have applied full reverse thrust and utilized every means possible to bring the aircraft rapidly to a halt, to allow a fast evacuation of the plane. Instead, he chose to taxi ponderously off the runway, trapping the passengers in the fume-filled aircraft for another 180 seconds. He then continued to run the engines for over three minutes, preventing rescue crews from taking action, and the cabin crew from initiating the evacuation.

As fire swept through the landed Tristar rescuers could only watch helplessly. It was clear that the rescue crews lacked adequate training, equipment, and leadership to deal with such an event. Nearly 30 minutes after the aircraft had returned to Riyadh, they were finally able to cut their way into the wreckage. All 301 people on board were dead, among them 15 babies. In the history of air disasters, it was a costly and needless waste of life almost without equal.

TENERIFE, CANARY ISLANDS, SPAIN

APRIL 25, 1980

Above: A Dan-Air Boeing 727 similar to the ill-fated aircraft that crashed in Tenerife carrying British tourists to the Canary Islands.

Tragedy struck the holiday island of Tenerife yet again on April 25, 1980, this time claiming the lives of 146 people flying in a Dan-Air Boeing 727. The accident highlighted the necessity for a clear and concise system of communication between aircrew and ground controllers. Both the KLM–Pan Am collision in 1975, and the destruction of the Dan-Air Boeing 727 were in part attributable to inadequacies in this communication.

Spanish air traffic control came under intense scrutiny during the late 1970s and early 1980s. Many aviation analysts questioned whether the organization possessed either the professionalism or the equipment

to cope with the volume of traffic that was operating out of their national airports.

However, given the particular circumstances and the actions of the Dan-Air crew in the minutes prior to this tragedy, it would be unjust to lay too much blame at the door of Spanish air traffic control operatives.

It seems that in this case, mutual incomprehension and casual oversights by human beings combined to create a human catastrophe.

In the late morning of April 25, 146 people (138 passengers and 8 flight and cabin crew) boarded the Dan-Air charter service from Manchester, England, bound for the beaches of Tenerife. The flight was uneventful up until the point where the aircraft passed into the jurisdiction of approach controllers at Los Rodeos airport in Tenerife. The weather conditions in the locality consisted of broken overcast, with a cloud base down to 3000 feet (1000m).

At this point, inaccurate navigation by the crew had resulted in the aircraft being nearly a mile (1.5km) to the east of its correct flight path. The approach controller, who was operating without the benefit of radar, was unclear as to the position of the aircraft and subsequently cleared the crew to descend to 6000 feet (1828m) and to enter a holding pattern in anticipation of a landing slot. The failure of the crew to report their position accurately after crossing over the VOR navigational aid (a ground-based electronic navigational aid) further aggravated the situation.

Further confusion arose when the controller instructed the crew to turn to the left, which in fact would have taken the aircraft away from Los Rodeos airport. This apparently irrational order disorientated the crew for vital seconds. As a result, the crew failed to enter the race track holding pattern over Los Rodeos, and continued flying towards the mountainous area to the southwest of the airport.

The aircraft was flying in cloud at a height of 5500 feet (1660m), when the ground proximity warning system sounded, and almost immediately deactivated as the 727 flew over a valley. The pilot, clearly confused about the location of the high ground, made a rapid evasive movement to the right at this time, but nearly 30 seconds later the ground proximity warning system sounded again. This time there was no opportunity for evasive action. The aircraft slammed into a mountainside and disintegrated, instantly killing all those on board.

Below: Rescue workers search for debris from the Boeing 727 that crashed in the woods south of Los Rodeos airport, Tenerife, killing all 146 passengers and crew.

LEONARDO DA VINCI, LA SPEZIA

JULY 3, 1980

The 33,340-gross ton *Leonardo da Vinci* was an elegant and luxurious liner that was built to replace the ill-fated *Andrea Doria*, which went to the bottom in 1956 following a collision with another vessel. The *Leonardo da Vinci* was built by Ansaldo SpA of Genoa and was launched in early December 1958.

The vessel was opulently fitted out – it had five swimming pools and 30 public rooms designed in modern Italian style – and it was an immediate hit with passengers when it began sailing between Genoa and New York. As well as plying the transatlantic route, the *Leonardo da Vinci* also sailed as a cruise ship in many of the world's oceans. In July 1977 the ship was transferred to Italian Line Cruises International to be used for short cruises between Florida's Port Everglades and Nassau in the Bahamas, but the scheme was far from successful. The *Leonardo da Vinci* was returned to La Spezia, laid up, and offered for sale in 1978.

The liner never sailed again. On July 3, 1980, a fire broke out in the ship's chapel and engulfed the whole vessel. The La Spezia fire service was unable to bring the fire under control and the *Leonardo da Vinci* was towed out of the harbor to burn itself out. The ship eventually heeled over and capsized. The *Leonardo* was raised in March 1981 but was too badly damaged for repair and was scrapped the following year.

Below: After being consumed by flames for three days, the burned-out hulk of the Leonardo da Vinci *lies on its side in La Spezia harbor.*

ALEXANDRA PALACE, LONDON, ENGLAND

JULY 10, 1980

Much of the magnificent Alexandra Palace in north London was destroyed by a fire which started accidentally in the organ loft shortly after workmen had begun preliminary restoration of the famous Henry Willis instrument inside it. No one was seriously injured in the blaze, but millions of pounds' worth of damage were caused and for some time it was uncertain whether the site could be rebuilt or if it would have to be flattened and covered with grass.

Despite its name, Alexandra Palace was not a royal residence but an exhibition building. Occupying seven acres (2.8 hectares), it featured Europe's largest hall – a distinction it retained right up until the day it was destroyed – a concert room, reading room, theater, and offices. It had burned down once before, just over a fortnight after it originally opened in 1873, when a red-hot coal fell from a workman's brazier, but it was rebuilt and was back in operation by 1875.

Alexandra Palace was never entirely successful as an exhibition center as most businesses continued to prefer the Crystal Palace in south London. But the extensive potential office space in Alexandra Palace and the fact that it was built on some of the highest ground in London made it attractive to the BBC's fledgling television service. In 1936 the company acquired part of it for television studios and a transmitter, and it was from here, on August 26 the same year, that the world's first television broadcast was made.

By the time of the fire, although the BBC's main television center had moved to west London, "Ally Pally" – as it was generally called – still contained the television studios of the Open University.

Left: Smoke pours from Alexandra Palace as firemen make desperate efforts to contain the blaze.

Right: The wreck of the great exhibition hall, which had been constructed in metal and glass in imitation of the Crystal Palace.

The fire was first spotted by a police patrol flying past in a helicopter on routine surveillance duty above the British capital. It spread quickly and destroyed at least 60 percent of the complex. More than 200 firemen rushed to the scene and fought the blaze, but much of the Palace was destroyed, along with sound equipment worth more than £250,000 ($400,000) which had recently been brought in for a planned jazz festival. About 350 BBC staff were evacuated.

The overall glass roof of the great exhibition hall collapsed in a series of explosions, buckling the metal supports and sending two million square feet (186 thousand sq m) of glass crashing to the floor. A pall of smoke rose high into the air and could be seen in central London, six miles (10km) away.

After enormous efforts to raise the finance, Alexandra Palace was partially restored and the new complex houses exhibition and banqueting halls and an ice rink.

RODO'S AND EL HUECO'S CLUBS, LONDON, ENGLAND

AUGUST 16, 1980

Thirty-seven people were burned to death or asphyxiated when two clubs in a single building in central London were firebombed by an arsonist with a grudge.

These establishments were not trendy night spots but illegal after-hours drinking clubs in Denmark Place, a sordid alleyway on the edge of Soho. Almost every-

thing about them was sleazy and even their names were unreliable – they were also known variously as the Colombian Club, the Spanish Club, and Victor's.

The man responsible for the deaths was John "Gypsy" Thompson. He torched the clubs because he thought he had been overcharged for a drink. Angry at having had to pay £5 ($8) for a rum and coke, he took a minicab to a nearby garage and bought a gallon (4.5 liters) of gasoline. At 0300 hours he went back to the building, poured the gasoline through the letter box, and threw a lighted match after it.

The flammable vapor caught light at once. There was a large explosion and a fireball rushed up the narrow staircase to the clubs on the two upper floors of the three-story building. (The ground floor was used to store the barrows of traders who operated in the street during the day.) Within seconds the whole interior was engulfed in flames. People sitting at the bar on the first floor had no time to react and fell dead still clutching their drinks. On the top floor, some of the customers in El Hueco's had a moment's warning of the danger and rushed into the tiny kitchen at the rear, from which they tried to escape through a small window two-feet (0.6m) square. A few managed to get out and slide to safety down the sloping roof below, but most were overwhelmed by the flames. The shambles in the kitchen showed that people had tried to climb over each other in a desperate fight to escape. As a police fire investigator put it: "There wouldn't have been room for them all to have stood normally in that small space."

The fire brigade brought the blaze under control at 0330 hours, but by then the building was gutted and part of the slate roof had collapsed. They found 24 victims in Rodo's and 13 in El Hueco's on the top floor. Most of the bodies were so badly mutilated that it was difficult to tell even what sex they were. This, together with the fact that most of the drinkers were illegal immigrants, made identification immensely difficult. Everyone was eventually named, partly because some of the victims' passports contained their owners' fingerprints.

Although Thompson thought no one had seen him firebomb the clubs, he was identified by eyewitnesses. He was afterwards convicted of murder and sent to prison for life.

Left: Firemen at work near the door (on the right) to the two clubs, which were on the upper floors of the building.

PRINSENDAM, PACIFIC
OCTOBER 4, 1980

Above: *A US Coast Guard helicopter hovers over the crippled* Prinsendam *as it is towed through the waters of the Gulf of Alaska.*

Right: *A lifeboat full of survivors from the burning* Prinsendam. *All on board were rescued safely and taken to coastal settlements in Alaska.*

The fate of the *Prinsendam* shows how the total loss of a ship need not be accompanied by loss of life. When fire broke out on the vessel the captain and crew followed the correct procedures, with the result that all the passengers and crew were evacuated and saved.

Built by Rotterdam's De Merwede Shipyards in 1972, the *Prinsendam*'s career as a cruise liner lasted only until 1980. A forewarning of its eventual fate came in April 1973 when a fire destroyed the ship's passenger accommodation and much of its superstructure.

When repairs had been completed, the *Prinsendam* was sent out to the Far East. The intention was for the ship to cruise around Indonesia, but the business was far from profitable and the ship was switched to the Vancouver and Singapore cruise trade.

The ship sailed from Vancouver during the winter, but switched to Singapore during the summer. In late 1980 the *Prinsendam* was based at Vancouver. In early October the vessel took on board over 300 passengers for a 29-day cruise. Soon after midnight on October 4, as the *Prinsendam* made its way through the Gulf of Alaska, a fire started in one of its main engines. The crew acted quickly, sealing off the area, and dowsing the flames with carbon dioxide, which should have extinguished the flames. However, when they moved back into the area of the fire, it rapidly became clear that the flames were still spreading.

Recognizing the severity of the situation, the captain sent out a distress call shortly after 0100 hours. The response was swift – helicopters, the US Coast Guard, and the supertanker *Williamsburgh* hastened to the stricken ship. Meanwhile the *Prinsendam*'s crew redoubled their efforts to halt the spread of the flames. It was to no avail. The fire led to a failure in the ship's electrical circuits, and water pressure, essential to the firefighters' efforts, plummeted. The ship's master, Captain Cornelius Wabeke, had no other option. At 0515 hours he gave the order to abandon ship.

The evacuation proceeded without panic and the passengers made their way safely into the lifeboats. The captain and 50 volunteers elected to stay on board to fight the fire, but little could be done. By mid-afternoon, the situation was beyond saving and Wabeke and his team also abandoned ship. All the passengers and crew were picked up by the rescue vessels and landed at a number of coastal settlements in Alaska.

The ship was now empty and was drifting. On October 7 a tug, the *Commodore Straits*, attempted to tow the *Prinsendam* back to Portland. The tug was able to get a line on the *Prinsendam* but the weather was worsening. The vessel was slowly but surely listing and taking on water from the rough seas. The fires on deck burned themselves out on October 10 but those below raged unabated. The ship's list was also worsening and on the morning of October 11, it was clear it could not survive. The *Commodore Straits* cut its line and the *Prinsendam* sank at 0835 hours.

Below: The scene on board the Prinsendam *after the passengers had been safely evacuated. Charred and blistered paintwork is evidence of the intense heat of the fire that had raged through the early hours of the morning.*

GRAND HOTEL, LAS VEGAS, USA

NOVEMBER 21, 1980

One hundred people were burned to death or asphyxiated and more than 600 injured, many of them seriously, when fire swept through the 26-story MGM Grand Hotel in Las Vegas, Nevada, USA. The blaze sent noxious smoke billowing through most of the 2000 rooms where many of the 3500 guests were asleep.

The fire started at about 0700 hours in the kitchen of the ground floor restaurant. Within moments, the adjacent casino was ablaze throughout its 147-yard (134m) length. The flames knocked out the hotel's electrical circuits so rapidly that the automatic fire alarms did not go off. Later, when guests telephoned reception for instructions they could not get through because the switchboard had also been put out of action. After the fire was discovered, no one thought to set off the manually operated alarms – even if they had been activated they would have been of little benefit because only three floors of the hotel had their own sprinkler systems. Although this may seem shocking with hindsight, it did not violate Las Vegas fire regulations at the time.

The smoke rose unchecked up the building through stair wells and lift shafts. As it spread it fed off plastic furnishings and became ever more lethal to inhale.

Below: The Las Vegas firefighting services tackle the fire at the Grand Hotel. The fire started early in the morning on the ground floor, and destroyed the electrical circuits so that the fire alarms did not ring.

Right: *At the height of the fire a great cloud of black smoke and poisonous fumes from the burning furnishings rose above the hotel. A helicopter hovers on the left of the picture, seeking an opportunity to rescue hotel guests from the roof or balconies.*

When it reached the upper floors, many of the guests remained uncertain what to do: despite the apparent danger, the absence of any alarm engendered a false sense of security. As the fumes intensified the guests at last recognized the danger they were in. The ensuing scenes of panic were described by a man who watched the tragedy from a hotel across the street as "sheer bedlam." Smoke had now enveloped nearly all the building, and guests waved sheets or threw chairs and tables through the windows of their rooms in a desperate attempt to get fresh air to breathe.

Six helicopters from a nearby US Air Force base were scrambled and rescued many of the guests by picking them off the balconies of their rooms and from the roof. Meanwhile, police helicopters circled the hotel pleading with guests through bullhorns not to jump. One woman did jump and was killed.

At street level, those who had been near the main exit or had managed to reach it swarmed out of the hotel in frightened droves. One witness described how the casino girls and dealers rushed away from the inferno "with cash drawers in their hands stuffing chips in their pockets."

Most of those who died were trapped on the upper floors, far beyond the reach of the fire brigade's ladders, which extended only nine storys. When the flames were eventually brought under control, firemen began to evacuate the building floor by floor, using window-cleaners' platforms to winch people to safety.

Left: *Firemen encourage a hotel guest to escape down one of their ladders, which unfortunately only extended as far as the ninth story. Other guests were rescued in window-cleaners' platforms.*

REINA DEL MAR, MEDITERRANEAN

MAY 28, 1981

The *Reina del Mar* began life in 1951 as the *Ocean Monarch*, and served with a number of passenger lines, changing its name in 1967 to the *Varna*, before being sold to a Greek shipping line in 1978. But the ship was destroyed by fire before it ever put to sea for its final owners.

The ship that was first called the *Ocean Monarch* was built in 1951 by a British shipyard, Vickers Armstrong, to serve the route between New York and Bermuda for Furness Withy. As the *Varna*, it was next owned by a Bulgarian company that chartered it out as a cruise ship sailing out of Montreal in the early 1970s. It then had a brief career with Sovereign Cruises from 1973.

The *Varna* made just two cruises for Sovereign and was laid up until 1978, when it was bought by a Greek shipping line based in Pireaus. The new owners

decided to change the ship's name to *Rivera*. The company had big plans to use the *Rivera* as a cruise ship, but they were slow in getting off the ground. In 1981 the *Rivera* was renamed *Reina del Mar*, and this seemed to do the trick. The owners announced that the *Reina del Mar* would commence a series of Mediterranean cruises in 1981.

But first it was decided to renovate the vessel. During the work a fire broke out on May 28 in the ship's boiler room and quickly spread through the vessel, devastating the passenger accommodation. The ship was left in a dangerous condition – it was burned out and its superstructure had collapsed. As it was a hazard to other ships, the *Reina del Mar* was towed to safety close to the *Rasa Sayang*, another burned-out vessel. However, the *Reina del Mar* capsized and went to the bottom off the yard at Perama on May 31.

Above: The Reina del Mar *berthed at Southampton in May 1974, seven years before the fatal fire that sent it to the bottom.*

WASHINGTON, USA

JANUARY 13, 1982

On a bitterly cold January afternoon, in the depths of one of the worst American winters on record, 78 people died when the Air Florida Boeing 727 that was carrying them to the warmer climes of Fort Lauderdale and Tampa, in Florida, smashed into a bridge and plunged into the ice-covered waters of the Potomac River.

Only the tail of the aircraft remained above the surface, and four passengers and a stewardess who had been seated in the rear of the cabin were able to scramble free and cling to the wreckage. A Bell LongRanger helicopter of the US Park Police arrived on the scene within 20 minutes, and plucked them from the freezing water. Two pedestrians on the bridge bravely dived into the water to assist in a desperate search for survivors. One of them undoubtedly saved the life of a woman who had lapsed into unconsciousness and lost her grip on a rescue line.

It was later concluded that one of the primary factors in the crash was the presence of ice on the wings and in the engine intakes. Although the 737 had been de-iced (common procedure in cold weather aircraft operations) about 60 minutes before the accident, this allowed sufficient time for ice to build up on the wings again. This in turn seriously hampered the performance of the aircraft when it became airborne, and caused it to enter an unrecoverable stall.

Below: The Boeing 727 had just taken off from Washington National Airport when it crashed into a road bridge. Here rescue workers on the bank of the partly frozen Potomac River help survivors ashore.

New Orleans, USA

July 9, 1982

Right: A Pan American Boeing 727 similar to the one that crashed after taking off from Moisant International airport, New Orleans, on July 9, 1982.

The term "microburst" is probably unfamiliar to most of the millions of people who board commercial airliners every year, but it is a word that strikes fear into the hearts of many an experienced pilot. It is a weather phenomenon – most often found in stormy conditions – in which a column of air moves rapidly downward and mushrooms out in all directions on reaching the ground. An aircraft flying into a microburst can experience a rapid loss in altitude, and often there is little or nothing the crew can do to recover.

For many years the microburst was a completely unexplained phenomenon. The accident described here prompted an extensive investigation into microburst that led to the development of low level wind shear alert systems (LLWAS). Although not foolproof, these could help to prevent a repetition of similar incidents.

Most modern airports are now fitted with LLWAS that can detect the presence of microbursts. This equipment is positioned at the ends of runways, to provide air traffic controllers with information they can relay to any aircraft preparing to take off or land. Unfortunately, at the time when this incident occurred, the technology was very much in its infancy.

The southern United States are prone to sudden and spectacular summer thunderstorms, which rise over the Gulf of Mexico before sweeping along across the Mississippi Delta. Just such a storm blew in to the New Orleans area on the afternoon of July 9, 1982. At Moisant International airport, serving New Orleans, Pan American Flight 759 – a Boeing 727 – was being prepared for the second leg of a three-leg scheduled service from Miami, Florida, to San Diego, California.

With preparations completed, and the passengers embarked, the crew taxied to the end of Runway 10 in anticipation of take-off. It was raining heavily at the time, the cloud base was at around 4000 feet (1200m), and visibility was approaching two miles (3km).

The captain had plenty of experience in such conditions and had been warned by airport meteorological staff of the likelihood of wind shear only five minutes before he was cleared for take-off. The crew of the 727 also tried to predict the weather conditions they would encounter in their immediate flight path by viewing the

Left: The wreckage of the Pan Am 727 litters the ground around a house that caught fire after the aircraft crashed into a residential neighborhood, killing eight people on the ground.

aircraft's weather radar, but the heavy rain probably prevented them from observing the threatening storm cells (which could have given rise to a microburst) in their path.

Moments after take-off, the Boeing 727 flew into a microburst that tossed it about the sky. The first officer, who was piloting the aircraft, battled helplessly and in vain to arrest the airliner's descent. After striking a stand of trees 2400 feet (730m) from the end of the runway the 727 impacted the ground and disintegrated. All 153 people on board were killed, as well as eight people on the ground.

The inquiry could find no fault in the operating procedures of either the airline or the flight crew.

Above: Rescue workers survey the scene of total devastation created by the crashed Boeing.

SALANG PASS TUNNEL, AFGHAN

NOVEMBER 2 OR 3, 1982

At least 1000 and possibly as many as 2700 troops and civilians were burned to death or suffocated after fire broke out in a road tunnel through the mountainous Hindu Kush. The tragedy occurred during the USSR's long – and ultimately unsuccessful – military campaign (1979–1989) to take over Afghanistan.

The Salang Pass tunnel was the principal road link between Kabul, the Afghan capital, and the USSR. It was thus of great strategic importance to both sides in the conflict. The Soviet forces guarded it ferociously in order to keep their supply lines open and it was often the target of attacks by guerrillas of the Afghan Muslim resistance movement.

The Soviet-built tunnel was 1.7 miles (2.5km) long, 17 feet (5m) wide and 25 feet (7.5m) high; it was poorly ventilated and lay at an altitude of 11,000 feet (3300m).

The disaster began when a petrol tanker was involved in a head-on collision with the leading vehicle in a Soviet military convoy. The tanker burst into flames almost immediately and was destroyed along with about 30 other vehicles, including transport trucks and buses. Several hundred men were burned to death at once. Most of the victims, however, died from asphyxiation by inhaling exhaust fumes – this was because in winter, high in the mountains, it was intensely cold and vehicles further down the convoy – whose drivers may not even have realized what was happening – kept their engines running. This quickly created an unbearable build-up of poisonous gases in the confined area of the tunnel.

But the most important contributory factor to the enormous number of fatalities was that, on hearing the explosion, Soviet troops mistook it for an enemy attack and immediately closed the tunnel at both ends,

Below: A Soviet patrol guarding a mountain pass in Afghanistan. When the officer guarding the Salang Pass tunnel heard the explosion, he sealed off both ends of the tunnel, trapping everyone inside.

Above: The Hindu Kush mountain range, separating the USSR and Afghanistan, was traversed through the Salang Pass road tunnel.

trapping everyone inside. This meant that everyone inside the tunnel died. Four days after the accident, the bodies of Soviet victims were still being airlifted to Kabul, while the Afghan dead and injured were taken to Jellalabad near the Pakistan border.

One eyewitness said that shortly after the tragedy a Soviet investigator who had been sent to the scene had a violent public argument with the officer whose decision it had been to close the tunnel.

Other details of the tragedy are sketchy, because the movement of troops is always a closely guarded secret and in the former Soviet Union there was no uncensored news coverage in peacetime, let alone in time of war. Even the exact date of the Afghan tunnel disaster is unknown, although news of it had reached Western intelligence sources by November 9. It is thought that the disaster may have taken place during the late afternoon about one week previously.

STATUTO CINEMA, TURIN, ITALY

FEBRUARY 13, 1983

Right: A body is carried out of the Statuto Cinema after the fire that killed 74 people. The rear emergency exits were found to be locked, and the manager of the cinema was subsequently arrested.

Seventy-four people were killed when fire swept through the Statuto Cinema in the center of Turin, Italy. There were at first thought to be only 37 dead, but firemen later discovered the same number of bodies in the upper gallery and in the upstairs toilets, where they had been suffocated by smoke fumes.

The fire started in the stalls and spread quickly, engulfing seats with plastic covers from which deadly smoke billowed up into the gallery. There was at the time no clear legislation in Italy to prevent the use of upholstery which gave off toxic fumes when hot. This helps to explain why the smoke was so bad that firemen were still having to wear breathing apparatus several hours after the blaze had been brought under control. Although people from the stalls were killed – many in the stampede to escape – the survivors had all been sitting on the ground floor, close to where the fire broke out. Everyone upstairs died.

Police said that there was no evidence of arson and that the blaze had probably been started accidentally by one of three things: a burning cigarette end, an electrical short circuit, or a firework set off as a joke – the tragedy occurred on the Sunday two days before Shrove Tuesday and fireworks are always popular at this time of year as Italians celebrate the carnival at the beginning of Lent.

Whatever started the fire, it would almost certainly have caused fewer deaths if the attendants had been on duty at the rear emergency exits, as they should have been. In their absence, the back doors had been locked to prevent latecomers getting in without paying. This meant that nobody could get out that way, and when the fire brigade arrived they had to break their way in with axes.

Of the remaining members of the audience in the 1070-seat cinema, most escaped unscathed: only three people were injured. Nearly all the victims were young: the average age was only 25 and the youngest were a boy aged 11 and a girl aged seven.

Among the dead was Giacomo Fracchia, aged 20. He was a Cuirassier – one of the Italian Army's ceremonial guards – and that is one of the reasons why all those killed in the fire were offered an official funeral in Turin Cathedral after a funeral Mass celebrated by the Cardinal Archbishop of the city and attended by the President of Italy. This honor was angrily turned down by 15 of the other victims' families – many of whom were poor economic migrants from the south of the country – who condemned it as ostentatious political interference in private grief.

Below: Most of the victims of the fire were given an official funeral in Turin's cathedral attended by the President of Italy.

ASH WEDNESDAY, AUSTRALIA

FEBRUARY 16, 1983

I n many parts of Australia, bush fires are a recurring fact of life. But in February 1983 they became a multiple killer when 68 people died in a conflagration that spread across many parts of Victoria and South Australia. Adelaide, the South Australian capital, was covered by a pall of smoke blown in from other parts of the state. This happened at the start of Lent and as a result the tragedy became known as Ash Wednesday.

Ash Wednesday was not one fire but many fires which started at about the same time in different places and joined up to make one immense inferno. Most of them were accidental, an almost inevitable consequence of the heat and the dryness of the bush. Others, however, were started deliberately, and one 19-year-old man from Adelaide was later convicted of arson.

The fires broke out at the height of the summer, when temperatures were in the region of 100°F (38°C). South Australia was in the middle of its worst drought for a 100 years – many farmers there had not seen rain for more than 12 months and thousands of their sheep had had to be shot for lack of drinking water.

Once the flames started, the fires spread rapidly, fanned by northerly breezes which gusted at speeds of up to 50mph (80kph). Vast tracts of the two states were declared disaster areas with hundreds of homes destroyed and huge numbers of sheep and cattle burned to death. More than 20 people were killed in South Australia, most of them near Adelaide, while at least 10 others died in separate incidents in Victoria.

Police said that some of the victims had been trapped in cars as flames suddenly swept across a main road between Adelaide and the outlying hills. Twelve

Above: Firefighters were powerless to stem the blaze as strong winds fanned the flames into a terrifying conflagration.

firemen were killed, including three who died together in their truck as they were fighting one particular blaze. Ambulance services reported that 230 people in the southeast Adelaide region had been taken to hospital suffering from burns or from the effects of smoke inhalation. All soldiers in the area were called up for firefighting duties, but by the time they arrived many of the blazes raging through bush and eucalyptus trees were already out of control. Some people tried to save their homes by dousing their roofs with water, but the drought had necessitated water cuts and there was not enough water available for the purpose.

Melbourne airport had to be closed for 20 minutes as thick smoke swept in from fires to the south of the city. Residents in the Adelaide Hills suburb were forced to flee their wooden homes, dozens of which were destroyed by fire. One of the worst affected areas was around Lorne, 70 miles (112km) southwest of Melbourne. The coastal town was surrounded by fire and hundreds of people had to rush to the beach as flames swept out of the bush and into their homes. Four towns – Macedon, Melton, Riddell's Creek, and Woodend – in the Mount Macedon area about 50 miles (80km) northwest of Melbourne had to be evacuated.

Left: A householder inspects the remains of her home. Dozens of timber-built houses succumbed to the fire.

Left: The fires quickly spread out of control, destroying everything in their path and killing 68 people.

Right: A couple in front of their fire-ravaged house in Melbourne. The fires swept through much of South Australia.

SEA OF JAPAN, OFF SAKHALIN ISLAND

SEPTEMBER 1, 1983

During the Cold War between the US and the USSR there were many incidents involving incursions into enemy airspace that were never reported in the press. Both sides were unwilling to admit that they were conducting highly secret reconnaissance missions by overflying each other's territory. These missions resulted in a number of hostile engagements and established a situation of extreme tension in the air that resulted eventually in the tragedy of Korean Airlines Flight 007.

Below: A Korean Airlines Boeing 747 in flight. It was an airliner similar to this that was shot down by a Soviet jet fighter in September 1983.

On September 1, 1983, the cat and mouse game that had been played in the skies for decades resulted in the downing of a commercial airliner and the loss of 269 lives. A Korean Airlines Boeing 747, designated Flight 007, left JFK airport in New York at 0420 hours GMT

with 240 passengers and 29 airline staff on board. It was on a scheduled flight to Seoul, South Korea.

The aircraft landed to refuel at Anchorage in Alaska and took off again without incident. At this point the flight crew should have entered fresh data into the trio of inertial navigation systems on the flight deck. These work in conjunction with the autopilot to steer the aircraft on the correct course to its destination. However, at 1310 hours and only 10 minutes into the final leg of its journey the 747 began to deviate from its designated course, and onto one that ultimately led it to pass into Soviet airspace.

The Soviet Air Force was placed on full alert, and interceptor aircraft were scrambled into the area the Korean airliner was overflying. This area was the Kamchatka peninsula, one of the most sensitive Soviet

Right: Japanese fishermen attempting to retrieve pieces of the 747's wreckage with fishing nets.

Below: In Seoul, South Korea, bereaved relatives weep as they burn incense before an altar at a memorial service for the 269 people killed when the Boeing 747 was shot down.

Pacific Fleet ballistic missile submarine bases. This fact makes the actions of the Soviet defence forces understandable, if not justifiable.

The failure of the Soviet planes to locate the airliner in the darkness did nothing to alleviate the growing tension on the ground. As the Boeing 747 began to approach Sakhalin Island, another highly sensitive Soviet military base in the Sea of Japan, the tension on the ground reached fever pitch.

At 1805 hours GMT, as the dawn rose over the Pacific, the pilot of a Sukhoi Su-15 jet fighter verified on radio that he had made visual contact with the errant jetliner (whose passengers were asleep and totally unaware of the seriousness of their situation). Later examination of voice-data recordings revealed that at no point did the Soviet pilot indicate that the aircraft was in fact a commercial passenger plane.

The Soviet pilot shadowed the airliner for a period of approximately 20 minutes, and attempted to draw the attention of the flight crew with standard IFF (identification friend or foe) code procedures. As a last ditch effort, he fired a burst of cannon fire ahead of the flight deck. After failing to elicit any response he launched two air-to-air missiles, one of which impacted on the airliner's left wing. The crippled aircraft, flying at over 30,000 feet (9000m), experienced a rapid cabin decompression and went into a steep dive. No one stood a chance of surviving the impact, which occurred roughly 52 miles (86km) off Sakhalin Island.

The Soviet authorities did little to aid either the process of recovering the wreckage or that of establishing how such a tragedy could have occurred. Instead they blamed the incident on a combination of US military aggression and Japanese air traffic control incompetence. It was later confirmed that a US Air Force RC-135 reconnaissance aircraft, similar in appearance to the Boeing 747, was operating in the area that night.

Two theories have been put forward as to why the aircraft deviated from its designated route. The first was that the crew failed to set the autopilot correctly on leaving Anchorage. The second was that the flight crew had inputted data that was incorrect by 10° into the inertial navigation system. This last theory would accord with the heading that the aircraft was on when it was shot down.

There is no doubt, however, that procedures and regulations in all quarters were given less than the required attention. And the fact that the Soviet pilot and defence forces failed to make sufficient information available in the aftermath of the tragedy led to widespread condemnation.

Below: The Korean Airlines Boeing 747 was shot down by a Soviet Sukhoi Su-15 jet fighter similar to the one shown here.

ALCALA 20 DISCO, MADRID, SPAIN

DECEMBER 17, 1983

Eighty-three young people were killed in the early hours of the morning when fire ripped through Alcalà 20, an underground discothèque near Puerta del Sol in central Madrid, Spain. Most of the dead were either asphyxiated or trampled under foot. The fire began at about 0445 hours when the disco was about to close and the rock music had been turned off. In contrast to many fatal fires in public places, the club was not overcrowded – there were only about 600 people in a building licensed to hold 900. But, as the subsequent investigation disclosed, this was one of the few things that could be said in favor of the place – in almost every other particular, Alcalà 20 was a fire prevention officer's worst nightmare.

One survivor said that the first he knew about the fire was when he saw "a flame leap out of the stage curtains. Within seconds there were fumes, hardly letting you breathe. That is when the stampede started, with hundreds of people trying to escape. I tried to grab the hands of some girls I saw being trampled underfoot, but it was like an unstoppable avalanche."

Another survivor spoke of how he had groped his way through the smoke looking for the ground floor exit but had been unable to find it because it was so badly signposted. He finally managed to clamber up to street level through a ventilator shaft, the protective glass at the top of which had apparently been smashed by someone escaping ahead of him. Many of those who did reach the exit ended up in a crush because the doors were too small to release all the people inside in time – it was just inside the main entrance that the rescue services eventually found the greatest number of trampled bodies.

Situated in the refurbished basement of an old theater, the Alcalà 20 discothèque had opened only three months previously. The conversion of the premises had been approved by all the relevant Spanish authorities, including the Madrid College of Architects and inspectors from the Industry Ministry.

After the tragedy, in response to criticism of locating such a public venue below ground level, Madrid's Civil Governor, Señor José Rodriguez, said: "Most

Left: The fire-damaged entrance to the Alcalà disco. In the stampede to get out, the doors proved to be too small and many trampled bodies were found near the exit.

Right: The scene in the disco after the fire. Lights on the dance floor short-circuited and within minutes the underground disco was an inferno.

discothèques all over the world so far as I know are underground because of the noise of the music."

It was subsequently discovered that the fire had probably been started by an electrical short circuit in the lights on the dance floor. The poisonous fumes had been given off from plastic curtains, wall hangings, and upholstery. Although there was a brief period during which the fire might possibly have been brought under control, the club's 10 fire extinguishers were either too difficult to operate or did not work at all.

Right: Firemen search through the debris of the gutted dance hall, looking for bodies of victims killed by the fire.

YORK MINSTER, YORK, ENGLAND

JULY 9, 1984

Above: *Although the fire gutted the south transept of the cathedral, the medieval stained glass windows escaped untouched, including the beautiful circular rose window.*

A bolt of lightning during an electrical storm in the summer of 1984 struck one of the pinnacles on top of York Minster and started a fire that destroyed nearly all the south transept, causing damage worth over £1 million ($1.6 million). Despite the enormous destruction, many people were grateful that the disaster had not been much worse – no one was hurt, and the greater part of Europe's largest Gothic cathedral remained unscathed.

In the early evening of July 8, an electrical storm raged over York and local people noticed that flashes of lightning appeared to play particularly about the roof of the Minster. The blaze was first spotted at 0230 hours in the early hours of the next day. The cathedral was fully protected against lightning and had numerous fire detectors throughout the concourse and lightning conductors on its pinnacles. Nevertheless, the sheer speed of the spread of the flames defeated all attempts to extinguish them, even after the rapid arrival of fire engines from nearby Northallerton.

The blaze was intensified by the cathedral's old, dry timbers and lead roof. It took almost three hours to bring the fire under control, and by then the ancient roof beams and plaster vaults of the transept (the part of the cathedral running at right angles to the nave) had been reduced to a smoldering mass of debris on the floor. In the morning light it was discovered that the high gable on the south wall of the transept was no longer supported by the roof.

Despite the horrendous damage, there was universal joy that the fire had been put out before it had reached the Minster's world-renowned collection of stained glass, particularly the unique 15th-century rose window. Many of the cathedral's treasures were saved by teams of clergy who ran relays and braved fire and showers of molten lead to salvage priceless artefacts. Only when the roof beams gave way did they give up. At the height of the blaze, a Minster clergyman and his wife dashed into the cathedral and rescued valuable items from the high altar before they were told firmly by firemen that they must leave.

Gossip at the time linked the fire to the recent consecration as Bishop of Durham of the Right Reverend David Jenkins, a controversial liberal clergyman – some suspected arson by protesters against his appointment, while others thought it was divine intervention, an Act of God. Police quickly ruled out the former, while of the latter Dr Robert Runcie, Archbishop of Canterbury – who was in York at the time for a General Synod – said "it seems absolutely miraculous that the fire was confined to the transept. York Minster has risen before and it will rise again." He also quoted the words of the Chief Fire Officer: "The Lord was on our side as we battled with those flames, and every man in my brigade knew they were doing something special by saving York Minster."

Below: In the aftermath of the fire that destroyed the roof of the south transept, experts check on the damage.

BRADFORD CITY FOOTBALL STADIUM, ENGLAND

MAY 11, 1985

Fifty-six people were burned to death, 70 more detained in the hospital with severe burns, and a further 211 supporters and police were injured when a timber stand caught fire at Bradford City's Valley Parade ground during an end-of-season game. The fire broke out just before half time beneath wooden tip-up seats three rows from the back of Block G during the final home league match of the season against Lincoln City.

The game was fairly well attended, because Bradford had just ensured promotion from the old English Third Division, and many fans had come to celebrate. The fire began amid rubbish that had accumulated beneath the stand over a long period and had never been swept up. The felt and wood roof, which was tinder-dry, fed the flames that rapidly engulfed the structure. When the fire was first noticed, there were a few tongues of flame licking the base of a row of seats. Within two minutes it had spread the entire length of the stand, moving faster than grown men could run.

Most of the victims were trapped in the stand itself. Many of the men, women, and young children who died were so badly mutilated that they could be identified only from dental records. The first to perish were those who tried to reach the back of the stand; of those who fled onto the pitch, most were saved. At least 15 bodies were found in a walkway four feet (1.3m) wide which ran along the entire length of the back of the stand.

A dozen bodies were found in clusters of two or three lying against six of the exits. They had been crushed to death as they desperately attempted to crawl out under turnstiles which had been locked to prevent latecomers getting in without paying. To make matters worse, there were no fire extinguishers in this part of the ground: they had been removed and stored in a room in the clubhouse because during previous games they been set off and used as missiles by unruly fans. The Chief Fire Officer of West Yorkshire, interviewed in *The Times*, said that as far as he knew there had never been a fire inspection at the ground because under English law the fire brigade was not empowered to carry one out on private property.

Right: An aerial view of Bradford football stadium showing the burning stand. The fire was fanned by a draft of air from the back when someone opened a door.

Above: The stand quickly became an inferno. Most of the people who jumped onto the pitch escaped, but those who tried to get out at the back found the turnstiles locked.

One spectator said: "We all thought it would be out in a few minutes. Everyone was telling us not to panic." Another survivor reported: "We could see the flames creeping along the bottom of the seats, then it seemed that someone opened a door at the back to let people out. There was a northerly breeze, which created a tunnel of wind, and it was like a furnace. People panicked and were rushing onto the pitch. I went across to help people into the ambulances and there were men with their hair burned off and their faces burned. There were children walking around with burns on their hands."

Left: Fans watch the stand burn from the comparative safety of the pitch. In all 56 people died in the disaster and many more were severely burned, or otherwise injured, and kept in hospital.

THE MOVE FIRE, PHILADELPHIA, USA

MAY 15, 1985

At least 11 people – four of them children – were killed and more than 250 left homeless as fire swept through a row of 61 houses in Philadelphia, Pennsylvania, USA. The disaster occurred when police dropped a bomb on the headquarters of Move, a radical group which described itself as "anti-society" and claimed to have renounced modern technology and everything to do with the US establishment. But one thing was never clear: was the tragedy an accident or had the police taken a deliberate decision to use extreme force, no matter what the consequences might be to nearby property and human life?

The bombing ended a 24-hour siege at a house in Osage Avenue during which police had fought the armed radicals in an attempt to evict them from the building. "We did not create any fire," said Police Commissioner Sambor. "To the best of our knowledge, the Move members had spread flammable material in their compound and in the neighboring area."

Sambor's claim that the police had not known the danger until it was too late was backed by Wilson Goode, the Mayor of Philadelphia, who insisted that the fire had been an accident: he said that the bomb had been meant to knock out Move's rooftop fortifications and open the way for the use of tear gas or water

Above: Clouds of black smoke rise over a residential street in Philadelphia after the police bombed a house belonging to Move, a radical, anti-social group.

Right: The fire spread rapidly through a row of 61 houses, leaving more than 250 people homeless, and 11 people dead.

Below: An armed policeman keeps watch on the fire created by the bombing.

to clear out the group without the danger of any loss of life. Explosives would not have been used, he said, if the authorities had known that there was flammable material in the house.

Asked why the fire brigade had not begun to tackle the blaze until it had been raging for at least an hour, city officials explained that the firemen were afraid they might be shot at – but the fire commissioner had already gone on record saying that leaving the fire to burn "worked to the city's advantage."

Many important questions remained unanswered: why had the bomb been dropped after Move had apparently announced over the loudspeaker system – with which they regularly harangued local residents – that they had gasoline in the house? Why had the police attacked when they knew that there were children inside the building? One member of the group – a woman known as Ramona Africa (all Move members took the surname Africa) – was arrested after escaping from the burned-out house and charged with making terrorist threats. In court, she asked the judge: "When are you going to charge Wilson Goode with murder?"

As the fire raged uncontrolled, many other houses in the street were destroyed and 250 people unconnected with Move were left homeless. Despite being promised that all the buildings would be rebuilt by Christmas, the residents sued the City of Philadelphia, its mayor and police chief for 10 million dollars. Later that year, under intense media scrutiny, Police Commissioner Sambor resigned from his post.

ATLANTIC OCEAN, OFF IRELAND

JUNE 23, 1985

When an Air-India Boeing 747 blew up in the air off the coast of Ireland in 1985, all 329 people aboard were killed. This major air disaster was the result of a series of mistakes and security lapses that had made sabotage possible.

The bomb that killed 329 people was almost certainly planted by a Sikh extremist in retaliation for the massacre by the Indian Army of Sikh worshippers at the Golden Temple, Amritsar, in the Punjab region of northern India.

The device was hidden in a suitcase and was probably placed on a Canadian Pacific aircraft designated as Flight CP60 at Vancouver. The man who made the reservation and checked in the suitcase requested that his luggage be transferred to Flight 182, the doomed Air-India 747, at Toronto – the destination of Flight CP60. A passenger with his surname also checked in at the same desk that day, with a single item of luggage, which was placed on Flight CP003 to Tokyo, Japan.

At Toronto, all the luggage being transferred from the Canadian Pacific aircraft to the Air-India Boeing should have been scrutinized with X-ray equipment and by random searches. Instead, due to the X-ray equipment being out of order, the luggage was screened by a hand-held electronic device that can, in theory, detect the presence of explosives. However, in this case it obviously failed to do so. Another lapse in security, and one that contravened airline policy, was

Below: An Air-India Boeing 747 flying over mountainous terrain.

Above: Part of the wreckage from the crashed Air-India Boeing 747 being brought ashore at Foynes, Ireland, from a Norwegian freighter. All wreckage from the aircraft was taken to Cork for investigation.

Right: Officials from Air-India examining some of the wreckage on board the Irish naval vessel Aisling moored at Cork, Ireland.

the failure of airline staff to match passengers to their luggage. An unaccompanied piece of luggage was allowed on the plane.

The saboteur never boarded the aircraft. At 0715 hours in the morning of June 23, in clear skies above the North Atlantic, his murderous device exploded in the forward cargo hold of the Air-India 747. The rapid decompression that follows when the pressurized fuselage of an aircraft flying at height is holed killed many passengers, and probably also caused the flight crew to collapse unconscious at the controls as they battled to save the stricken jumbo. Wreckage was strewn over a large area, and 95 percent of it sank without a trace into the Atlantic. Only 132 bodies were retrieved from the sea – there were no survivors.

Approximately an hour before the Air-India Boeing exploded, a second device contained in the luggage the man had placed on Flight CP003 exploded prematurely while being transferred to Air-India Flight 301 at Narita airport in Tokyo. Two people were killed. But for an incorrectly set timer in this second device, the death toll would almost certainly have been much higher.

In October 2000, after years of investigation, Canadian police arrested two members of Canada's Sikh community for conspiring to bomb the aircraft.

MOUNT OSUTAKA, JAPAN

AUGUST 12, 1985

The crash of the Japan Airlines Boeing 747 into Mount Osutaka on August 12, 1986, was one of the most harrowing, as well as one of the worst, air disasters on record.

Japan Airlines Flight 123 was on a scheduled service from Haneda International airport, serving Tokyo, to Osaka 250 miles (402km) to the west. Many of the passengers were returning home to be with their families for the traditional festival of Bon. The routine flight would normally have taken a little under an hour.

After taking off at 1812 hours, the Boeing 747 had levelled off to its cruising height of 24,000 feet (7300m) when a loud explosion was heard at the rear of the aircraft. Pressure then began to drop rapidly in the passenger cabin and the emergency oxygen masks dropped from the roof.

As the captain made a Mayday signal to Tokyo Control, one of the cabin crew reported that part of the aft fuselage was missing. The control wires to the rear control surfaces in the 747 run along this section of the fuselage and when the flight crew tried to maneuver the aircraft they found that the controls had been rendered useless. The giant aircraft began to turn ominously northward toward the snow-capped peak of Mount Fujiama.

Although the controls had been irrevocably damaged, the flight crew were able to use engine power to steer the aircraft, and despite the total loss of hydraulic power they managed to lower the undercarriage using the back-up system. At about 1844 hours the jumbo began to descend, and after circling perilously over the city of Otsuki righted itself and flew on toward the mountainous region around Takasaki. Despite the considerable efforts of the crew the aircraft crashed at an altitude of about 4800 feet (1463m) into the side of Mount Osutaka.

Rescue crews did not reach the remote mountainous region until 0900 hours the next morning, and found a scene of almost complete devastation. A stewardess, a 12-year-old boy, an eight-year-old girl, and her mother were the only survivors.

Left: A scene of appalling devastation greeted the members of the Japanese Self Defense forces who began a search of the area using a helicopter on the day following the crash.

MANCHESTER INTERNATIONAL AIRPORT, ENGLAND

AUGUST 22, 1985

The year 1985 was particularly costly in human terms for commercial air carriers. Six major incidents during those 12 months cost over 1300 people their lives. One of the most horrific of these incidents was the loss of the British Airtours Boeing Advanced 737 at Manchester International airport.

The Greek islands enjoyed a significant increase in tourist traffic during the 1980s, and British Airtours, a subsidiary of British Airways, carried many hundreds of thousands of these tourists from Britain. In the early morning of August 22, 1985, a full complement of 130 passengers, and two infants in arms, boarded a British Airtours 737 at Manchester, bound for a vacation on the Greek island of Corfu.

While the passengers stowed overhead baggage, the cabin crew prepared for a busy flight. On the flight deck, preflight checks passed routinely, and the aircraft proceeded to the holding area just off Runway 24. Cleared for take-off, the captain released the brakes and the aircraft began to accelerate smoothly down the centerline. As the 737 speeded through 140mph (225km/h), the flight crew heard a loud thump. Believing that a tire had failed they immediately aborted the take-off run, and informed the control tower of their predicament.

However, what the flight crew had heard was the port engine partially disintegrating. Parts of the engine casing ruptured the fuel tank next to the engine, and as the aircraft began its emergency deceleration, aviation fuel gushed over the red-hot powerplant and ignited.

Left: Police and firemen inspecting the wreck of the British Airtours Boeing 737 which lost an engine and burst into flames during take-off at Manchester airport.

Above: Despite the efforts of the rescue services, 55 people lost their lives in the fire that engulfed the aircraft.

Right: Aviation inspectors check the port engine of the crashed 737.

This fire was not immediately indicated on the flight deck, where the crew were still under the illusion that a tire had burst. The flight crew then used the power of the engine reversers to arrest the progress of the jetliner, which served literally to fan the flames. When the aircraft finally turned off the runway and ground to a halt, the blaze was already intense. Aviation fuel spilled out of the wing tank and formed a flaming lake on the concrete. To further hamper the evacuation, the prevailing wind then fanned the flames toward the aircraft, burning into the passenger cabin within half a minute.

The scene inside the aircraft was chaotic, and as dark clouds of toxic smoke billowed into the cabin, the crew lost precious seconds struggling with a jammed door. Many people were overcome by smoke inhalation as they struggled in the dark and confusion toward emergency exits, and bodies jammed the narrow aisle. The sheer number of passengers, and the fact that two of the exits were engulfed in flames, further hampered the evacuation. Only about 60 seconds after the aircraft ground to a halt, the rear fuselage collapsed. Although the emergency services arrived on the scene with commendable speed, they were unfortunately unable to save 55 of the passengers, almost all of them in the rear cabin. Dozens of the survivors suffered injuries, many of them serious.

The report of the British Air Accident Investigation Branch revealed that the compression chamber of the Pratt and Witney turbofan had cracked and then partially disintegrated due to thermal fatigue. This is caused by continual heating and cooling of metal parts. The engine fitted to the British Airtours aircraft had previously been repaired for a crack in the same area, but the repair was of poor quality.

ARGENTON-SUR-CREUSE, FRANCE

AUGUST 31, 1985

The summer of 1985 was not a good time for the French national railroads. In the space of two months, three serious incidents shook the faith of the French public in their state transport system and led to the resignation of the president of the railroads. This incident, which left 43 people dead and a further 38 injured, involved a train scheduled to make the southerly journey from Paris to Port Bou on the country's Mediterranean coast close to the Spanish frontier and a postal train from Brive, south of Limoges, to Paris.

The crash took place at night on a section of track that was being overhauled. The driver of the Paris–Port Bou train received a warning in his cab to reduce his speed as he approached the section of track being upgraded. His train, which was advancing at nearly 90mph (144km/h), braked to 65mph (104km/h) after receiving the normal warning regarding the curve in the track. However, a second warning to reduce his speed to 20mph (32km/h) was misinterpreted. The driver believed this second warning to be nothing more than a reminder to keep to the 65mph (104km/h) limit on the curved section. Consequently, he continued south at this regulation speed, only applying the brakes when the danger of negotiating the curve became clear at the last moment.

His failure to recognize the second order to reduce speed over the track led to the derailment of the train. The driver, sensing danger at the approach of the Brive–Paris train, then attempted to warn his opposite number by flashing his lights. His frantic, last-minute efforts to prevent a collision were to no avail, however. The driver of the post train was unable to brake in the short distance available to him and smashed into the derailment at high speed.

Right: Rescuers crowd around the two trains involved in the incident at Argenton-sur-Creuse in central France. A derailment caused by excessive speed was followed by a collision.

NEWFOUNDLAND, CANADA

DECEMBER 12, 1985

Right: The wreck of the Arrow Air DC-8 lies on a wooded hillside after crashing half a mile from Gander International airport.

Ice always spells trouble for aircraft. Even a very thin sheet over the lifting and control surface can be potentially hazardous. In this case, it may have been the cause of a tragic crash.

Almost all of the passengers on the Arrow Air DC-8 scheduled flight of December 12, 1985, were US soldiers returning home to Kentucky after service in the Middle East.

After taking off from Cairo airport, the jetliner flew to Gander International airport in Newfoundland, which was in the depths of a Canadian winter. Sleet probably started to accumulate on the wings of the DC-8 during the stopover, and then froze onto the leading edge and upper surfaces of the wings. The DC-8 was already heavily burdened with the troops and their equipment; and reports subsequently showed that the crew had in fact underestimated the necessary take-off speed and "nose-up" angle required for the weight of the aircraft.

The DC-8 took off in the predawn darkness at 0645 hours, struggled to a height of only 120 feet (40m), and then began – ominously – to descend. Early morning commuters on the Trans-Canada highway were stunned to see the stricken aircraft pass only a few feet over their cars, before crashing into a wooded hillside barely half a mile from the end of the runway. Flames engulfed the aircraft, killing everyone on board.

The board of inquiry was unable to reach any firm conclusions as to the cause of the crash.

HINTON, ALBERTA, CANADA

FEBRUARY 8, 1986

Below: Sulfur and pipes lie scattered around the derailed and crushed remains of the trains involved in the crash. Intense heat from fires caused by the impact left much wreckage unrecognizable.

Twenty-nine people died in this collision, but the death toll could have been a lot higher given the immediate aftermath of the event. A pair of Canadian National trains were involved in the incident. One consisted of a large freight train pulling more than 100 wagons; the other was a passenger train. The crash was held to be the fault of the crew of the freight train, who failed to wait in a siding located in the foothills of the Rocky Mountains in western Alberta before the single-track section ahead of them was cleared of on-coming traffic.

The freight train continued its journey at high speed and smashed into the passenger train traveling in the opposite direction. So great was the closing speed of the two trains that both their locomotives exploded at the moment of impact. The leading coaches of the passenger train were left as a barely recognizable tangle of wreckage by the smash and their metal was melted and twisted by the subsequent fireball that engulfed them.

To make matters much worse for the rescue workers, the load of sulfur being carried by the freight train also caught fire, covering the immediate area with thick smoke and the poisonous fumes of sulfur dioxide. The more seriously injured passengers were evacuated by helicopter to the town of Hinton some 10 miles (16km) from the scene of the collision, while buses were driven to the accident site to take others to the city of Edmonton.

MARAVATIO, MEXICO

MARCH 31, 1986

One only has to observe an airliner landing to understand the brutal treatment that the main undercarriage is subjected to. Tires are also subject to great stresses and changes in temperature during every take-off or landing run. Despite this, and because of the expense of aircraft tires, they are often "retreaded," a process which in the past has proved to be fallible.

The most likely cause of this crash of a Boeing Advanced 727 operated by Compania Mexicana de Aviacion was the combination of poorly maintained brakes and an incorrectly manufactured tire. It brought a tragic end to a 10-year accident-free record for the Mexican carrier.

During its take-off run at Benito Juarez airport near Mexico City the left main gear on the Mexicana Boeing Advanced 727 (Flight 940) started to bind. Unknown to the crew, the drag that this caused raised the temperature of the brakes well beyond the safe limit. When they were retracted, the brakes would have been glowing red, and in the enclosed undercarriage bay would have raised the temperature of the tires to a dangerous level.

The crew had levelled off to their cruising height of 31,000 feet (9400m) for the short 400-mile (640-km) westbound transit to Puerto Vallarta when an explosion blew out a large section of the port wing, damaging vital fuel and hydraulic lines and electrical cables. At this height this caused an explosive cabin decompression. The crew made a terse emergency report before the aircraft crashed into mountains near Maravatio, 100 miles (161km) to the northwest of Mexico City. All 159 passengers and eight crew members were killed.

Right: The Boeing 727 crashed into a wooded, mountainous area, making the job of the rescue teams even more difficult. Here rescue workers carry away the body of one of the 167 victims.

HAMPTON COURT PALACE, LONDON, ENGLAND

MARCH 31, 1986

Above: The fire at Hampton Court Palace started in one of the "grace and favor" apartments on the top floor.

Far right: Firefighters on the roof damp down the epicenter of the fire.

A fire that devastated the Cartoon Gallery of Hampton Court Palace, the historic English royal residence on the outskirts of London, killed a general's widow and caused damage worth more than £1 million ($1.6 million).

The fire broke out upstairs in the Palace's south wing, designed by Sir Christopher Wren, one of the greatest English architects. The wing housed the Cartoon Gallery, named for the Raphael drawings that had been brought to England by King Charles I and were displayed there until the mid-19th century when they were removed to the Victoria and Albert Museum.

The upper floor of this wing was residential, a "grace and favor" apartment occupied by Lady Gale,

aged 86, the widow of General Sir Richard Gale, who in the 1950s had been commander-in-chief of the British Army of the Rhine. Lady Gale had been in the habit of reading by candlelight, and this is thought to have been the cause of the disaster. Palace staff later revealed that they had been concerned for some time at Lady Gale's increasing inability to fend for herself. One aide was reported as saying: "She was in the habit of taking a drink at night by candlelight and falling asleep sometimes with the candle still burning. Some of the ladies had feared a fire might start."

The fire probably broke out at about 0230 hours but went undetected until 0530 hours, when the alarms went off. Although Hampton Court Palace had fire extinguishers, there was no sprinkler system because

it was feared it might go off accidentally and damage some of the priceless furnishings and paintings.

The emergency services were summoned at 0545, by which time a huge pall of smoke could be seen more than two miles (3km) away. Six other elderly residents – the widows of generals, diplomats, and former colonial service officials – were woken immediately and moved out of their top-floor apartments to safety. Over 120 firemen from all over London and Surrey were mustered to fight the blaze. Tight security protecting the Palace treasures meant that they had to smash their way with pickaxes through the Cartoon Gallery's barred and reinforced doors.

Not long afterwards the roof of the building collapsed and crashed onto the Cartoon Gallery 40 feet (12m) below. The smoke was appallingly thick. There was an enormous amount of timber and other debris falling all the time and it was impossible to see or hear anything through the glare and noise of the inferno.

Firefighters rescued everything that was portable, but they were too late to save two valuable paintings, along with much furniture and oak panelling. Two other famous paintings – "The Field of the Cloth of Gold" and a portrait of King Henry VIII – were blackened by smoke and drenched with water from the firemen's hoses, but they were salvageable and were taken at once to the restorers.

Above: Despite the efforts of the firefighters, much of the south wing of the Palace was destroyed.

Right: Although the Tudor part of the Palace (seen on the left here) escaped the fire, half of Sir Christopher Wren's 17th-century south wing (in the foreground) was devastated, including its ancient roof timbers.

LOS ANGELES, USA

AUGUST 31, 1986

Recreational flying is extremely popular in southern California, and the pilots of light aircraft passionately defend their right to use the airspace above the state. However, few light aircraft operating out of the numerous airports in the area carry sophisticated avionics equipment and the Piper Archer single engine monoplane involved in this accident was no exception. The transponder it carried was a non-encoding version, which does not register the altitude of the aircraft on the radar screen of an air traffic controller.

In the minutes before this incident, the pilot of the Piper strayed unwittingly into the path of a commercial airliner. Although Los Angeles air traffic control was aware of this intrusion, in the moments before the impact the duty controller's attention was distracted by the appearance on his radar scope of yet another light aircraft that had strayed into the restricted airspace around the terminal.

An Aeromexico McDonnell Douglas DC-9, on the final leg of a scheduled service from Mexico City to Los Angeles, was making its final approach to the runway when it collided with the Piper.

The tailplane of the DC-9 knifed into the the smaller aircraft, killing the pilot and his two passengers – his wife and daughter. The force of the impact tore off part of the Mexican aircraft's tailplane, causing it to roll over and enter a terminal 45° dive.

Falling on a residential area in the Los Angeles suburbs, the Mexican DC-9 airliner exploded in a fireball. All 64 people on board were killed, together with another 15 on the ground. Wreckage from the Piper monoplane crashed down into a schoolyard a few hundred yards (meters) away, but miraculously there were no further casualties.

Right: The scene on the ground after a Mexican DC-9 hit a private monoplane and crashed into a largely residential area in a Los Angeles suburb, destroying at least 10 houses.

SANDOZ FACTORY, BASEL, SWITZERLAND

NOVEMBER 1, 1986

A relatively small fire in a chemical factory on the banks of the River Rhine near Basel, Switzerland, resulted indirectly in one of the most serious cases of environmental pollution the world has ever seen.

The incident occurred at the Schweizerhalle works of Sandoz, the Swiss chemical manufacturing conglomerate. Fire broke out in Warehouse 956, an open-plan hall measuring 295 by 164 feet (90 by 50m) and 32 feet (10m) from floor to ceiling.

Here, a consignment of Prussian blue artists' pigment was loaded onto a palette and then shrinkwrapped by heating plastic sheets with a blow torch. The flames were supposed to be kept at a distance of at least 12 inches (30cm) from the plastic, but it is believed that the torch may have been held too close, and the package overheated. This almost certainly happened during the day shift, but Prussian blue may glow smokelessly for up to 12 hours before it ignites. It was not until 0019 hours that a night worker first noticed a fire. At almost exactly the same time, a Basel police traffic patrol spotted flames shooting from the roof of the warehouse and alerted the fire brigade. The works' fire crew arrived at 0022 and within minutes they were joined by back-up teams from Basel and a fireboat on the Rhine. Within the hour, 160 firemen were in action at the scene.

The fire was put out shortly before 0500 hours and the firemen then turned their attention to cooling down the smoking ruins. But by this time large quantities of toxic chemicals had already been released in the form of smoke – these included carbon monoxide, sulfur, phosphorus, nitrogen, and carbon.

Worst of all was dioxin, a lethal by-product of the manufacture of herbicides which – under its other name, Agent Orange – was used as a defoliant in the Vietnam War.

Below: When fire broke out in the Sandoz chemical factory in Basel it released large quantities of toxic fumes into the atmosphere.

Right: Firemen extinguishing the last of the fire in the factory. An enormous volume of water was used to put out the fire, and this water subsequently drained into the Rhine River, polluting it with many poisons.

Below: One of the divers employed to clean up the Rhine is scrubbed clean after surfacing. Large vacuum hoses were used to suck up the contaminated mud at the bottom of the river.

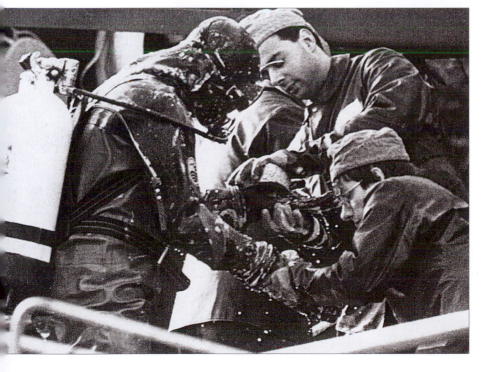

Within days, the Rhine had turned an intense red color. This was at first attributed to the dyes that had been stored in Warehouse 956, but it later became apparent that about 3.3 million gallons (15,000 cubic meters) of water used to douse the flames had been contaminated with insecticides before being drained into the river via the sewer network. The poison spread downstream, and eventually more than 155 miles (250km) of the Rhine suffered severe ecological damage and many forms of wildlife were affected.

Because of fears about the poison being communicated to humans through the food chain, both the Swiss and German governments immediately banned all fishing in the Rhine. Thousands of square yards (meters) of the river bottom were then cleaned by the removal of the upper layer of silt with electrosuction equipment – more than 2200 lbs (1000kg) of toxic waste were recovered in this way. Although the Rhine was declared safe on July 1, 1987, the Sandoz company faced claims for compensation amounting to 100 million Swiss francs. In addition to numerous individual claimants, compensation was also sought by several German cities downstream of the disaster.

OFF SUMBURGH, SHETLAND ISLES

NOVEMBER 6, 1986

Right: British Aviation Minister Michael Spicer inspects pieces of the crashed Chinook helicopter that were recovered from the sea off Sumburgh.

During the 1980s British International Helicopters operated regular services under contract with the various oil companies to oil rigs and drilling platforms in the North Sea for the benefit of oil company personnel. Both Chinook and Sikorsky helicopters regularly made the journey in poor weather conditions – four accidents, two of them fatal, were recorded between 1986 and 1990. The first was the worst ever commercial helicopter accident.

The Chinook helicopter took off at approximately 1115 hours on November 6, 1986, carrying 42 workers back to Sumburgh airport near Lerwick at the end of their three-week stint on a rig in the Brent oil field.

Three miles out to sea from the airport, in strong winds and rain, the rear set of rotors on the twin rotor-craft struck the forward blades. The rear blades, and the transmission driving them, then broke away and the helicopter fell from an altitude of about 500 feet (150m) into the freezing waters of the North Sea. In all, 45 people on board died of the trauma they suffered as a result. The captain and a single passenger were plucked from the sea by a helicopter that arrived on the scene only shortly after the crash.

When wreckage recovered from the sea was analyzed, it became clear that one of the main gears driving the rotor head had cracked through fatigue, possibly exacerbated by the corrosion associated with salt water operations.

HERALD OF FREE ENTERPRISE, ZEEBRUGGE

MARCH 6, 1987

The cross-Channel ferry service was, even before the opening of the Channel Tunnel rail link, a highly competitive business. Townsend Thoresen, one of the big operators, took over the P&O Group in April 1987 and as part of the deal gained a number of RO-RO (roll-on, roll-off) ferries, including the 7591-ton *Herald of Free Enterprise*.

Roll-on, roll-off ferries carry coaches, lorries, and cars across the English Channel, and are fitted with massive doors at both bow and stern to speed up the loading and unloading processes. A quick turnaround is vital to profitability. The *Herald of Free Enterprise* usually worked between Dover and Calais but was transferred

to the Zeebrugge run early in March to take the place of another ferry that was undergoing maintenance. On March 6, the *Herald* left Dover at 1130 hours. The crossing, which went without a hitch, was completed by late afternoon. There was meant to be a two-hour stay at Zeebrugge, loading passengers and vehicles. The *Herald* was due to depart on its return leg at about 1730 hours, but was delayed because of time it took to process a large, unseasonable number of passengers taking advantage of a cheap-crossing promotion.

Zeebrugge harbor is somewhat confined and ferries have to maneuver carefully to reach open water. The *Herald* had to reverse stern first into a side-dock and then, bow-first, head out into the Channel. The *Herald*

Left: When the car ferry Herald of Free Enterprise *rolled over just outside Zeebrugge harbor, many passengers were trapped inside and drowned.*

Left: Giant barge cranes pull the Herald of Free Enterprise *upright before the ferry is towed away.*

accomplished the first stage of the maneuver but disaster struck as it moved forward. The main bow door, giving entrance to the car deck, had been left open. As the ship moved forward at around 18 knots, tons of water were scooped up and entered the deck. As the ship rolled, the water also rolled from side to side, setting up a lateral motion that sent the vessel over on to its port side in less than a minute. It came to rest on a sandbank just outside the port's outer breakwater.

Chaos reigned inside the stricken vessel. Passengers below deck panicked as they fought to get out before being drowned. The lights went out and water continued to flood into the vessel. Belgian rescue services were quickly on the scene of the disaster and began to rescue the crew and passengers. Some 408 were brought out alive almost immediately, along with 50 bodies. The final death toll reached nearly 200. The *Herald* was eventually righted by a Dutch salvage company, Smit International, and then sailed to the Far East, where it was scrapped.

The board of inquiry found Townsend Thoresen wholly responsible for the disaster. In particular, it noted that several of the company's captains had already expressed their disquiet over the regulations for closing doors, and that there was no means for a captain to check for himself that the doors were shut. The inquiry stated that the company's management was "infected with the disease of sloppiness from top to bottom." Ferries were henceforth to be fitted with an indicator light to show whether doors were open or closed.

Left: A frogman in a dinghy inspects the open bow doors of the capsized ferry. Netting was hung over the opening to prevent cargo escaping.

OKECIE AIRPORT, POLAND

MAY 9, 1987

Even the failure of the smallest component in a commercial jetliner can spell disaster. In this tragedy, a worn ball-bearing began a series of events that led to the death of 188 people.

Filled almost to capacity, an Ilyushin IL-62 jetliner of Polskie Linie Lotnieze (LOT Polish Airlines) took off on May 9, 1987, from Okecie airport near Warsaw on a transatlantic service to New York City.

Many of those on board were on their way to see relatives. As the aircraft climbed to its cruising altitude, at 31,000 feet (9400m), the port inner engine rapidly disintegrated, wrecking the port outer engine. Parts of both engines smashed into the passenger cabin, as well as damaging the fin and elevator. The pilot regained control, but because of the damage to the electrical system, there was no indication of the fire that had started in the hold.

Although the pilot managed to guide the stricken aircraft back to Okecie for a landing approach, the fire damage in the rear of the aircraft caused a total loss of control. The aircraft cut a wide swathe through a forest as it crashed three miles (5km) short of the runway, with the loss of all those on board.

Below: A fireman searching for remains of the Ilyushin airliner that crashed into a forest near Warsaw killing all those on board.

GASOLINE TANKER, HERBORN, GERMANY

JULY 7, 1987

A fully loaded gasoline tanker smashed into a busy ice-cream parlor in the German town of Herborn, killing 50 people and injuring a further 25. Fuel leaked from the truck and spread right across the town center – as a result, when the gasoline caught light a couple of minutes later, the whole place was quickly turned into a firebomb. In the immediate aftermath of the disaster, the municipal authorities declared a state of emergency and evacuated 20,000 nearby flats and houses. Although most people were allowed back into their homes five hours later, those who lived closest to the blast had to be accommodated overnight in local schools.

The tanker, which was carrying 7000 gallons (32,000 liters) of gasoline, had left the nearby Frankfurt-Ruhr autobahn with overheated brakes shortly before it crashed into the building housing an ice-cream parlor and pizza restaurant at about 2050 hours. The death toll might have been much greater, had it not been for the fact that the pizza restaurant was closed.

The tanker caught fire on impact with the wall. Many of the dead were trapped inside the restaurant, which was quickly razed to the ground. A little later, the flames set off gas pipeline explosions in eight adjacent houses, three of which were completely destroyed. Ambulances, fire engines, police, and rescue services raced to the scene. After tending to the

Below: When a tanker went out of control and crashed into an ice-cream parlor the resulting explosion set off a fire that affected most of the center of Herborn.

Above: *After the initial impact, gas explosions in adjoining houses caused further devastation.*

most seriously injured, their main concern was the possibility of further explosions. Gasoline from the tanker had flooded into the underground sewage system and the surrounding air was thick with inflammable and toxic fumes: almost any part of Herborn was now a potential inferno.

Meanwhile, the roads into the town had become blocked by hundreds of cars full of ghoulish sightseers who had come to look at the scene of carnage. As a result, precious time was lost while helicopter ambulances were scrambled to airlift the injured to hospital. Amazingly, the tanker driver himself was only slightly hurt. This was because he had had enough time after the impact and before the explosion to jump out of his cab. The time lag between the crash and the explosion was confirmed by bystanders. One, Roger

Schmidt, described how he had rushed into the street after hearing the tanker smash into the corner of the building. "A couple of minutes later, there was a massive blast," he said. "I think the water in the sewage system exploded with the gasoline that had poured into it. Manhole covers flew into the air."

Another eyewitness, Hagen Puthner, aged 18, reported that he had heard the blast from his home where he had been having a meal with a friend. "There were flames and the earth was burning. People were crying for help. We tried to get a ladder but the flames were too high."

'There is hardly a brick left standing where the restaurant was," said a police spokesman, "It is a catastrophe. There have never been so many deaths here before. We are not used to this."

DETROIT, MICHIGAN, USA

AUGUST 16, 1987

The sole survivor of the Northwest Airlines Flight 255 of August 16, 1987, was a four-year-old girl. Tragically, she lost her other three immediate relatives in the accident, which robbed another 153 of their lives.

The US National Transportation Safety Board, which investigates all air accidents in the country, highlighted as the cause of the crash the lack of communication between the captain and his co-pilot while they were conducting take-off procedures at Wayne County airport at Detroit.

Specifically, the first officer failed to extend the flaps while the aircraft was taxiing just before it was cleared to take-off. The captain also showed a dereliction of duty by failing to complete preflight and take-off checks according to airline procedure.

Without either extended flaps or slats (retractable lifting surfaces on the leading edge of the wing) the McDonnell Douglas DC-9 Super 82 – and indeed any modern commercial transport aircraft – has a greatly reduced climbing capability.

After a ground run of about 6800 feet (2073m) the Northwest Airlines DC-9 lifted off the runway. If the aircraft had been properly configured, it would have been climbing through 600 feet (182m) when it was 4500 feet (1350m) further on its course. In fact, due to the fatal oversight of the flight crew, the aircraft had struggled to an altitude of barely 41 feet (12m). Only 14 seconds into the flight to Phoenix, Arizona, the port (left) wing clipped a string of lamp-posts and a building, and the aircraft plummeted onto a road about half a mile from the end of the runway.

The two occupants of the car that the aircraft also hit were killed instantaneously. As for the occupants of the aircraft, when the emergency teams arrived at the scene of devastation they were amazed to find that one person had survived the crash – a four-year-old girl.

Below: The scene on the highway where the Northwest Airlines DC-9 crashed soon after take-off.

JAKARTA, INDONESIA

OCTOBER 19, 1987

Above: *One of the factors behind the high casualty list in this disaster was the number of passengers who rode on the roofs and hung from the sides of the trains involved.*

ndonesian railroads have suffered many serious incidents since World War II. In April 1963, an express bound for Bandung from Jakarta left the track and one of its carriages was violently derailed and plunged into a ravine, leaving 37 dead. This incident echoed a similar event in 1959, when an express traveling to Bandung from Bandjar dropped into a ravine following an incident that was blamed on sabotage. Close to 100 people died and 14 were injured.

In November 1993, a pair of trains was involved in a head-on collision at Depok, a suburb of Jakarta. Both were badly damaged and there were a number of casualties. The impact was so severe that the leading part of each train was smashed into a near-vertical position which hampered the search for survivors. Severe though these three incidents were, they are comparatively small-scale events when compared with the disaster that overtook a pair of overcrowded commuter trains involved in a head-on collision which took place in one of the capital's southern suburbs.

It was not unusual for Indonesian trains to be overloaded. People were often to be found sitting on the roofs of carriages or hanging on to their sides or ends. On this occasion, it was to prove a fateful and lethal decision that contributed to a casualty list that totalled more than 150 dead and over 300 injured, many severely.

The impact was so violent that many carriages were crushed almost beyond recognition. Rescue services took many hours to sift through the wreckage in search

of survivors and the dead, and lighting had to be brought up to the crash site as night fell. Scores of troops from nearby army bases were rushed to the accident scene to help in the painstaking rescue work, but even with their invaluable help, the last surviving victims to be recovered from the wreckage, two young boys, were not brought out until nearly 20 hours after the crash.

Once the crash site had been cleared of casualties, the investigators were able to go about their business of searching for clues as to the cause of the collision. Initial reports from the crash scene suggested that the fatal impact had been brought about by a signalling error, but it was unclear if this malfunction was the product of a simple technical failure in the system's signalling apparatus, or whether it was caused by simple human error.

The Indonesian Ministry of Communications indicated that there would be a thorough investigation of the signalling system employed by the state's railroads, but were unable to rule out human error as a possible cause of one of the country's worst accidents in recent memory.

Above: *Troops and rescue workers pull victims from the wreckage of the two commuter trains involved in the Jakarta collision.*

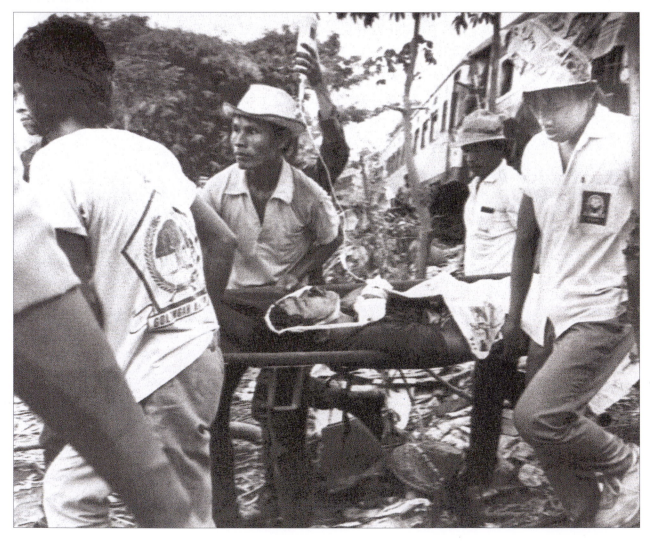

Right: *Medical staff rush a severely injured victim away from the scene of the crash.*

KING'S CROSS UNDERGROUND STATION, LONDON, ENGLAND

NOVEMBER 18, 1987

Above: Firemen inspect the damage after the fire at King's Cross underground station that killed 30 passengers at the height of the evening rush hour.

When a 48-year-old wooden escalator caught fire at the Piccadilly Line exit from King's Cross, London's busiest underground station, it was during the evening rush hour. In the resulting blaze 30 commuters were burned to death.

The blaze began when fluff under the escalator caught fire, possibly from a discarded match or cigarette end. The result was the type of fire known as a flashover – a searing wall of flame that rushed from the foot of the moving walkway to the booking hall upstairs, 50 feet (15m) below street level. PC Stephen Hanson, a London Transport policeman on duty that night, described it as "a shock wave of fire" that knocked him off his feet.

Assistant Chief Fire Officer Joe Kennedy, who was in charge of operations on the night, described the blaze as "one of the worst kind of situations that firefighters have to face. Because of the concrete tunnels, the heat generated by the fire was not released." Testimony to the ferocity of the blaze is that there were so few survivors: the fire took almost everyone it encountered.

All but one of the victims had been downstairs on or near the platforms when the fire was first detected. They had then tried to reach safety at ground level, but the fire – fanned by the draught of passing trains – pursued them up the escalator and killed them in the booking hall. The other fatal casualty was Station Officer Colin Townley of Red Watch at Soho Fire Station, who died fighting the blaze.

The inquiry into the King's Cross fire caused great disquiet about safety standards on the underground system. One revelation was that passengers had seen smoke coming from the same escalator earlier that day – this had been reported to London Underground staff, but the warnings had gone unheeded. Worse, there was no sprinkler system on the station.

There was further controversy about the amount of training London Underground staff were given to combat such emergencies. London Regional Transport, the administrative body in charge of the station, maintained that "all staff go through two and a half days' emergency training and have to take part in special local training when they move to a new station." Employees at King's Cross, however, said that they had been given no formal training whatsoever,

Above: *The remains of the escalators.*

Left: *The booking hall in the underground concourse.*

Far left: *The scene outside King's Cross.*

and the question of how to organize an evacuation in the event of a fire had never been raised, nor had there ever been a fire drill practice. One railman said that in 10 years' service at the station he had only been given a few minutes' instruction in using a fire extinguisher.

There were two important consequences of the King's Cross disaster – one was that wooden escalators were quickly phased out of the London Underground system; the other was that smoking was banned throughout the system, both on platforms and on trains.

One male victim of the King's Cross fire has never been identified. A decade after the event, an artist's impression of his reconstructed face was shown to the public in the hope he might yet be recognized.

DONA PAZ, PHILIPPINES

DECEMBER 20, 1987

The treacherous waters dividing the many islands that make up the Philippines have been the scene of many shipwrecks, but the incident involving the *Doña Paz*, a workhorse ferry used on the inter-island routes, ranks as one of the most horrific disasters ever to befall any vessel.

The *Doña Paz*'s last voyage took place at one of the busiest times of the year, when demand on the Philippine ferries is at its highest. The ferry was operating on the Leyte–Manila route and sailed from Tacloban for the capital packed with locals anxious to finish their preparations for the Christmas festival. There were so many passengers crowded on to the *Doña Paz* that many went unrecorded, so the lists of casualties after the disaster were far from accurate.

The first leg of the voyage was completed without any problems being encountered but disaster struck as the *Doña Paz* entered on the last third of its scheduled run and was just off the small island of Marinduque.

It was a dark, moonless night and many of the passsengers were sleeping on deck as best they could under the crowded conditions. Without any warning, the ferry suddenly collided with a tanker, the *Victor*. The impact was followed almost immediately by fire and the *Doña Paz* was quickly engulfed and sank. The surrounding sea was covered with blazing oil from the tanker and people in the water had little chance of survival. The few who did manage to escape the flames had to spend most of the night clinging to wreckage as the rescue effort did not get under way until daylight. Even then, the weather conspired to make rescue difficult. Heavy thunderstorms blanketed the area.

When the investigation was completed, it was estimated that the two vessels together were carrying 1556 passengers and crew, but this was almost certainly an underestimate. The figure was probably much higher. Only 30 people survived the collision and subsequent flames, and many of the bodies were never recovered from the sea.

Above: The inter-island ferry, the Doña Paz, *on one of its regular voyages.*

ARZAMAS, SOVIET UNION

JUNE 4, 1988

Above: The deep crater left by the detonation of 100 tons of industrial explosives. Scores of civilians waiting at a level crossing were killed in the blast.

This dramatic incident on the Russian rail network took place at Arzamas, a town 250 miles (400km) to the east of Moscow, and left an estimated 100 dead and 200 injured.

The cause of this massive casualty list was a huge explosion that also produced enormous damage to facilities in the vicinity of the detonation. The incident began as a freight train was approaching a level crossing at Arzamas. For some reason, possibly the unstable nature of their cargo, three of its wagons exploded.

The three wagons contained a massive amount of explosives destined to be used by various industries. The violence of the unexpected detonation can be gauged by the fact that the wagons and much of the track were obliterated, and a crater some 80 feet (25m) deep was gouged out of the earth.

The violence of the spectacular explosion flattened an estimated 150 dwellings around the level crossing and left a further 250 buildings suffering from various degrees of damage. The headquarters of the local branch of the Communist Party, more than a mile (1.6km) from the crossing, had its windows smashed by the shock waves and a vital gas pipeline close to the track needed extensive repair.

The long list of casualties was in part due to the line of vehicles waiting to cross the level crossing once the freight train had gone by. Most of their passengers were killed in the first few seconds after the explosion, engulfed by the blast.

GARE DE LYON, PARIS, FRANCE

JUNE 27, 1988

The legal proceedings that followed this incident involving two trains on the Paris underground system proved highly controversial. The prosecuting attorney at the trial, which began in late 1992, called for long sentences on conviction of charges of manslaughter to be imposed on a driver of the runaway train involved, a woman who had pulled a communication cord, a guard who had been late on duty the day of the crash, and a station supervisor who had apparently failed to order the evacuation of the public from the station.

The fatal collision involved a runaway electric-powered commuter train which hit a similar unit at the height of the French capital's evening rush hour. Nearly 60 passengers were killed and 32 serious injuries were also reported.

The woman who was later prosecuted apparently pulled the communication cord of the train from Melun as it was approaching Vert-de-Maisons, some five miles (8km) from the Gare de Lyon. She seemingly wanted to get home as quickly as possible and thought that the train was supposed to stop at her station, even though it was not scheduled to halt there. The driver had considerable difficulty resetting his train's brakes following the woman's intervention, but nevertheless

Above: The head of the crash investigation team visits the scene of the Gare de Lyon incident before the train wreckage was removed.

Right: Members of the French rescue services remove the dead and injured from the Gare de Lyon shortly after the crash.

Above: *Cutting gear is prepared to slice through the wreckage left by the crash so that it can be removed for further study.*

he continued on his way despite the fact that they were apparently barely operable.

To make up for nearly 30 minutes of delays caused by the unscheduled stop, the driver increased speed to nearly 60mph (96km/h). On approaching the Gare de Lyon, he realized that his brakes were still faulty and, believing that a secondary system would not operate effectively because of the wet conditions, attempted to avert disaster by sending an emergency message to the Gare de Lyon. However, he was little more than a mile (1.6km) out when this message was transmitted.

Unfortunately, other incidents intervened to increase the severity of the accident. A guard at the Gare de Lyon was late in beginning his shift on a second train, one bound for Melun, which was waiting in the station on the track adjacent to the side wall of the tunnel. The runaway train smashed through its

leading coach, creating a tangle of wreckage which took rescue teams most of the night to search through and free trapped survivors and remove the bodies of the dead.

At the subsequent trial, the women and the supervisor were acquitted, but the driver received a prison sentence of four years and the guard was given a suspended sentence of two years. Following a nationwide strike by railroad employees who greatly resented the court's decision, the guard's sentence was reduced to two months. The driver was held in jail for six months before being freed.

Following the Gare de Lyon tragedy and two other accidents a few weeks later at the Gare de l'Est, also in Paris, and at Toulouse in the southwest of the country, various new safety measures were introduced on the French rail network.

PERSIAN GULF

JULY 3, 1988

There are always risks attached to operating commercial transport aircraft in airspace adjacent to a war zone. In a war zone military personnel are operating in a heightened state of tension, and are forced to make split-second decisions that can – and do – lead to mistakes. The fate of Iran Air's Flight 655 demonstrates how a case of mistaken identity can end in tragedy.

The Persian Gulf during 1988 was an extremely hostile environment. The Iran–Iraq War had been raging since 1979, and had become a war of attrition. During the mid-1980s both sides had taken to attacking commercial shipping sailing in the Gulf, with the intention of denying vital oil supplies to the enemy.

To protect the oil it buys from Kuwait, the US committed part of the Fifth Fleet to the troubled region in May 1987 to escort vulnerable Kuwaiti tankers, which were forced to sail almost the entire length of the Gulf and through the narrow jaws of the Strait of Hormuz. A matter of days after the fleet was on station, the lives of 37 US sailors were lost when an Iraqi Air Force Mirage attacked the USS *Stark*, supposedly by accident. As a result of this tragedy US commanders were effectively given revised rules of engagement to enable them to defend US military personnel and equipment against acts of aggression.

A year on from the deployment, on Sunday July 3, 1988, the US frigate *Vincennes* became engaged in a surface battle with Iranian gunboats in the Strait of

Hormuz. Fully alerted to the possibility of an air strike, and with the *Stark* incident fresh in their minds, the crew of the *Vincennes* positively identified a radar echo emanating from the coastal area of Iran as an F-14 Tomcat, an aircraft which, ironically enough, the Americans had sold to Iran in the early 1980s. The aircraft was given a series of warnings on both the international air defence and military air distress radio frequencies – standard procedure in such situations.

Captain Rogers of the *Vincennes* was faced with a stark choice. His vessel was maneuvering violently while engaging with the Iranian ships. At the same time an aircraft confirmed as a military type by his air defence officer was apparently closing on his position. His decision to launch two surface-to-air missiles to counter the threat was perhaps a result of the pressures he was facing, and one that many people in his position would have made.

The passengers and crew of Flight 655 probably never knew what hit them. The aircraft wrongly identified as a fighter aircraft was in fact an Iranian jetliner – an Airbus A300. It had taken off from Bander Abbas, a city on the southwestern coast of Iran, at approximately 1015 hours with 278 passengers and 12 crew bound for Dubai, UAE, about 155 miles (250km) to the southwest. Only 20 minutes after take-off everyone on board was dead.

At a speed approaching 1900mph (3000km/h), the effect of a standard surface-to-air missile impacting on a wide-bodied jet would have been catastrophic. About 70 percent of the bodies and a large quantity of wreck-

Above: An Iran Air Airbus A300 similar to the one shot down in the Persian Gulf by the US frigate Vincennes.

Above: The USS Vincennes, *which was engaged in a surface battle with Iranian gunboats at the time of the incident.*

Above: *The radar screens on board the* Vincennes. *Information from these screens indicated to the captain that the Iran Air Airbus was a hostile aircraft.*

age were recovered from the waters of the Gulf. The large area over which the remains were spread suggests that the aircraft broke up in mid-air.

Inquiries were launched immediately by the US Navy, the Iranian government, and the independent International Civil Aviation Organization. The most crucial evidence came from tape recordings of the air defence radar on the *Vincennes*, which clearly showed that the Iranian Airbus had registered a transponder code identifying it as a civilian aircraft. A transponder emits a coded electronic signal that shows up on a radar screen as a series of numbers. A Mode II signal is used by military aircraft, Mode III by civil aircraft. Another anomaly was revealed by the data transcript, which showed that the airliner was climbing at the time of the impact, while the information given to the captain approximately 90 seconds before his fatal decision was that the aircraft was rapidly descending.

Further investigations revealed that as both civilian and military air fleets were operating out of Bander Abbas at the time, the Airbus had been confused with the transponder emission from an F-14 as it took off. Although the US Navy gave its full backing to Captain Rogers and his crew, the captain of a US naval vessel also involved in the surface battle on the day of the tragedy criticized Rogers and his officers for their openly aggressive actions.

PIPER ALPHA OIL RIG, NORTH SEA
JULY 6, 1988

A huge fire caused by two explosions on Piper Alpha, an offshore oil drilling rig in the North Sea about 120 miles east of Wick, Scotland, killed 151 men. Flames from the inferno leaped more than 200 feet (60m) into the air.

On the evening of Wednesday, July 6, 1988, the Aberdeen coast guard received a mayday call from the *Lowland Cavalier*, an oil rig support vessel in the North Sea. The message – "Explosion on the Piper Alpha" – was logged at 2158 hours.

The explosion had happened at 2131 hours, and the rescue teams were immediately puzzled why there had been such a long delay before the distress call had been put out. It was only when they reached the rig and

saw the extent of the devastation that they realized: there had been too much panic even to call for help.

Over the next two hours, there was a massive response by the British emergency services. Twelve RAF helicopters, a Nimrod search and rescue maritime patrol aircraft, six Royal Navy warships, a fisheries protection vessel, and an assortment of oil rig and coast guard vessels and commercial helicopters all made full speed to the scene of the disaster. The NATO Standing Naval Force Atlantic also became involved: this was a unique aspect of the operation.

In the minutes after the first blast, the men on board the rig itself had to make a fateful decision: whether to remain on the platform and wait for help to arrive, or to jump 40 feet (12m) into the water below, which was

Above: Two days after the explosion that claimed 151 lives, the Piper Alpha oil rig was still belching black smoke. In the background is a firefighting platform.

seething close to boiling point. The 70 men who survived this disaster all took their chance and threw themselves upon the mercy of the sea.

The first ship to sight the blazing platform was the *Tharos*. This support vessel – which had been patrolling nearby – had hospital accommodation for 22 as well as firefighting equipment. But before it could come within range of the stricken rig, a second explosion tore away the remains of the platform and killed anyone who had not already jumped overboard. The crew of the *Tharos* could do no more than watch as oilmen waved frantically for help from the helipad before the final blast – in the words of one onlooker – "just blew them away."

The *Tharos* picked the first survivors out of the water at 2210 hours, but further efforts to help were hampered by the intensity of the blaze, which forced the vessel to pull back a mile (1.6km) from the platform. The injured received emergency treatment on board the rescue ship.

The exact cause of the disaster was never conclusively established. One theory was that the explosions occurred after a sudden surge of gas escaped from a pocket in the reservoir beneath the North Sea oil bed. This might have been ignited by a chance spark on the platform. Survivors spoke of the squealing sound of escaping gas about 30 seconds before the first explosion: "a screaming like a banshee," as one of them described it.

In the aftermath of the tragedy, Piper Alpha was declared a safety hazard and was sunk. The Piper field, however, later resumed full production.

Below: About 70 percent of the Piper Alpha rig was destroyed by the explosion and fire. A diver support vessel lies close by as the rig still smolders.

RAMSTEIN, WEST GERMANY

AUGUST 28, 1988

Military aerobatic close formation flying is the ultimate test of skill and nerves, and a fiercely competitive arena in which pilots from the world's air forces try to show each other how good they are. They may also be putting the aircraft through its paces to impress potential customers. In short, they are the very public face of a country's military air forces.

The competition to win one of the coveted places in these teams, such as the British Red Arrows, or the Italian Frecce Tricolori, is intense, and only highly experienced pilots are considered for what many of their colleagues regard as a glamorous posting.

The Italian Air Force's Frecce Tricolori team was nearing the end of its display in front of a huge 100,000

strong crowd of German nationals and US service personnel at the NATO airbase at Ramstein in West Germany when this incident occurred.

At 1535 hours the team was performing a crossover maneuver when the solo lead pilot, Lieutenant-Colonel Ivo Nutallari, collided with the main group as they flew in close formation along the crowd line. Highly inflammable jet fuel and burning wreckage rained down on to the ground and cut a path through the crowd, killing 30 spectators, the three pilots involved in the collision, and injuring at least 60 other spectators. Many people were seriously burned and 13 spectators later died in hospital of their wounds. It was later assessed that the team had transgressed a strict air display rule by flying a maneuver that involved a solo aircraft flying directly at the crowd.

Below: The Frecce Tricolori team of Aermacchi MB 339 giving their display at the Ramstein air show just as the fatal incident occurs.

CLAPHAM JUNCTION, ENGLAND

DECEMBER 12, 1988

This early morning collision between two commuter trains packed with passengers heading back to work in London left 35 dead and nearly 70 suffering from serious injuries. A third train, one traveling down from London, then smashed into the wreckage caused by the initial smash. The incident began when the first train, traveling from Basingstoke in Hampshire to London's Waterloo station, was ordered to stop by a signal placed in a cutting a short distance to the west of Clapham Junction.

The signal turned from green to red as the train drew near and its driver applied the emergency brake. He then let the train coast to the next signal and reported the incident to the nearest signal box.

Left: Rescue workers toil in the narrow confines between the carriages of the trains involved in the crash at Clapham Junction. Dazed and injured passengers had to be lifted on to the embankment.

Above: A crane rips away a section from one of the crushed carriages as the process of clearing the line gets under way.

As he was telephoning, the Basingstoke–London train was hit in the rear by a train from the southwest of England traveling at 35mph (56km/h). It was at this point that the down train piled into the wreckage blocking the track. Emergency services rushed to the scene of the crash after being alerted by members of the public living nearby who had been stirred into action by the noise of the two collisions. They were confronted by a difficult task. The trains contained some 1500 passengers, many of whom were trapped in their carriages by twisted steelwork and seats that had been ripped from their fittings. They also had to negotiate a 10-foot (3m) retaining wall at the side of the up track against which many of the wrecked carriages had come to rest. The last casualty could not be removed from the site until some five hours after the crash had occurred.

An official investigation team led by a top lawyer, Anthony Holden, quickly discovered that the crash had been caused by wiring mistakes made when a signal was overhauled some two weeks before the crash. As part of the modernization of this busy line, the Waterloo Area Resignalling Scheme was given the go-ahead in late 1984, but the work progressed slowly so as to cause as little disruption to services as possible. A maintenance worker installing new wiring in a signal box at Clapham Junction failed to carry out the proper procedures. Instead of shortening the old wire at one end and binding it in new insulating tape and isolating it at its live end, he simply pushed the live end out of the way.

A day before the accident, further upgrade work was being carried out and during the course of this, the first wire moved back to its original position and completed the original connection. The result was that the signal in question was kept at green even when there was a train farther up the track. It was discovered that safety levels had been allowed to decline over the years and that staff had been working overly long hours to complete the work in hand.

Right: Fire crews and emergency workers continue their search through the wreckage of the trains smashed in the incident at Claphan.

LOCKERBIE, SCOTLAND

DECEMBER 21, 1988

At 2000 hours on December 21, 1988, members of the British Air Accidents Investigations Branch (AAIB) were racing to the small town of Lockerbie in Dumfrieshire, Scotland, to begin the harrowing task of piecing together the last moments of Pan American Flight 103 bound from London to New York.

The scene that greeted them in the sleepy town was one of widespread devastation. Wreckage from the Boeing 747-121 had plummeted into the southern edge of the town, gouging a huge crater more than 155 feet (47m) long in the residential area of Sherwood Crescent. A 60-foot (18-m) section of the rear fuselage had also crashed into a residential block about 2000 feet (600m) away. Debris and wreckage were scattered along two mains paths, and were recovered from 80 miles (130km) away down the east coast of England. There were no survivors among the 259 people on board the aircraft. The fact that there were only 11 fatalities among the residents of Lockerbie is nothing short of miraculous.

From the manner of the violent destruction of the aircraft it soon became clear to the investigative teams that the jumbo had been ripped apart by some sort of explosion while at altitude. Within a few days items of wreckage were recovered from which forensic scientists found clear evidence of the use of a sinister high

Below: The devastation in the small Scottish town of Lockerbie was caused by wreckage from the Pan Am Boeing 747 crashing into a residential block.

Right: Part of the wrecked cockpit of the crashed Boeing 747 being examined by investigators.

Below: The forward section of the fuselage of the crashed plane was found in a field about 2.5 miles (4km) east of Lockerbie.

explosive – Semtex – which is a favored tool of the terrorist. The investigation then developed into a murder hunt.

The Pan American Boeing had arrived at London Heathrow airport from San Francisco in the late morning of December 21. The aircraft was routinely cleaned and resupplied in preparation for its return flight across the Atlantic to New York's JFK airport. It was scheduled to depart at 1800 hours that evening.

Boarding the plane for the transatlantic flight were 49 passengers, together with their baggage, who had transferred from a flight that originated in Frankfurt, West Germany. In addition to these 49, another 194 passengers boarded, together with 16 members of the flight and cabin crew. At 1825 hours the Boeing took off and adopted a northwesterly course, heading through the center of Britain, and climbing to an altitude of 31,000 feet (9449m). At 1858 hours an air traffic controller at Shanwick (now Prestwick) Oceanic Area Control (the coordinating traffic control for transatlantic flights) transmitted a message to the Pan Am flight crew, confirming their clearance for the flight over the North Atlantic. When the message was not acknowledged, and the radar echo from Flight 103 broke up on their screens, officers at air traffic control began to fear the worst.

Residents in Lockerbie reported hearing a rumbling noise like thunder, which steadily grew louder like the

Above: A fireman searching the debris in the town caused by the impact of the disintegrating Boeing 747. Houses and cars were destroyed, and 11 residents were killed.

roar of a jet engine. Rushing to their windows, many of them saw the flaming hulk of the plane's fuselage, and the wing, crashing to the ground.

Approximately 90 percent of the wreckage was recovered – scattered over a wide area – during the course of the next 14 days. It was taken to a British Army base near Lockerbie. The forward section of the fuselage, which was recovered largely intact from a field about 2.5 miles (4km) east of the town, was reconstructed at the accident facility of the Royal Aircraft Establishment at Farnborough in England.

After many months of painstaking investigation, involving explosives experts, terrorist counter-intelligence experts, and members of the AAIB and the US National Transportation Safety Board, a preliminary report was published.

The detonation of an explosive device concealed in a Toshiba radio cassette recorder in the cargo hold had ruptured the pressurized hull of the aircraft, causing an explosive decompression, loss of control, and disintegration of the jetliner.

Embedded in a piece of luggage the investigators had found a tiny fragment of a Swiss timing device. The accusing eyes of the world turned on Libya when it was revealed that this same device had been sold to that country in 1985. The criminal investigation revealed that the device, contained in the baggage of a 50-year-old Libyan man, had been transferred from Air Malta Flight 180 at Frankfurt onto a Boeing 727, and then onto the Pan American flight.

Two Libyans accused of Lockerbie, Al-Amin Khalifa Fhimah and Abdel Baset al-Megrahi, were at the center of a political wrangle as to where they should stand trial for around 10 years, although it was finally agreed that they would be tried under Scottish jurisdiction but in the Netherlands. A former US Air Force base near Utrecht, Camp Zeist, was given temporary Scottish status to allow the trial to begin. The suspects were handed over in 1999 and the trial began the following year. In January 2001 al-Megrahi was sentenced to life imprisonment, but seemed likely to appeal, while Fhimah was found innocent and released.

LAVIA, HONG KONG

JANUARY 7, 1989

The process of refitting can be a dangerous operation for a ship, it seems. As with the *Reina del Mar*, the fire that ended the life of the *Lavia* broke out while the vessel was in harbor undergoing a refit programme.

The *Lavia* (first called the *Media*) was built in 1947 by John Brown and Company of Glasgow for the Cunard Line. Fitted with twin-screws and steam turbines, the *Media* was the first new passenger ship to be built for the transatlantic route since the end of World War II. Cunard had initially intended the vessel to be a cargo-carrier but had a change of heart. The *Media* was refitted with cabins to carry 250 first-class passengers.

By August 1947 the *Media* was ready to ply the seas between Liverpool and New York. However, the advent of fast jet travel between Europe and North America in the late 1950s was to put an end to the era of the great transatlantic liners. The *Media* was also too slow to compete with the new cargo-carriers being brought into service. Its career now began a downward spiral that would end in disaster.

In the early 1960s the *Media* was sold to the Italian Codegar Line and refitted in Genoa to carry 1320 passengers in tourist class. Renamed the *Flavia*, the vessel served the booming round-the-world cruise market and the lucrative trade transporting emigrants to Australia, but this upturn in its fortunes did not last.

Below: The large quantities of water used to put out the fire on the Lavia *caused it to capsize. It was later sold for scrap.*

Above: The Lavia *smoldering in Hong Kong harbor, with a fireboat alongside.*

After the Codegar Line was taken over by the Costa Line in 1968, the *Flavia* cruised between Miami and various Caribbean islands until 1982. After nearly 35 years in service around the world, the *Flavia* was showing its age and its turbines were becoming less and less efficient. A Chinese business based in Hong Kong, the Virtue Shipping Company, took on the *Flavia*, intending to convert it into a casino cruise ship. The ship was renamed the *Flavian* but did not prove popular and remained for most of the time at its moorings in Hong Kong harbor.

Perhaps hoping for a change of fortune, the Virtue Shipping Company changed the ship's name yet again, to the *Lavia*, in 1986, and decided to have the vessel refitted. The refitting work began but was never completed. In January 1989 a fire used by workmen got out of control and the flames spread through the ship's cabins. As the *Lavia* was in harbor, emergency vessels were soon on the scene. Four Hong Kong fireboats and more than 250 firemen attacked the flames, but the ship was fatally damaged, and the vast quantities of water that were sprayed over the fire caused the ship to heel over and capsize.

None of the 35 workers or nine crew on the vessel was injured, but the *Lavia*'s days were over. The ship was refloated, towed to Taiwan, and scrapped.

PUBAIL, BANGLADESH

JANUARY 15, 1989

Far right: Heavy lift equipment is used to remove parts of one of the derailed carriages from the track following the incident at Pubail which left over 150 dead and many hundreds injured.

There have been many catastrophic train incidents on the Indian sub-continent, but this collision must rate as one of the worst. In a matter of moments, 170 travelers were killed and over 400 injured, many seriously. The collision involved an express service from Chittagong, a major Bangladeshi port situated on the Bay of Bengal, and a post train heading north. Both trains were packed with commuters, who filled the seats and passage-ways, and perched on the carriage roofs. Many of the 2000 passengers on the train were religious pilgrims heading for Tongi to attend the largest annual gathering of Muslims outside Mecca in Saudi Arabia.

The collision caused carnage. Derailed carriages were overturned by the impact of the two trains and rolled down steep embankments into paddy fields adjacent to the tracks. Many of the passengers were crushed to death or suffered serious injuries as the carriages rolled over. Bangladeshi soldiers were rushed to the scene of the accident from a nearby camp, but their rescue efforts were hindered by the large crowds that had gathered to view the wreckage.

How then did the two trains and the passengers meet their fate? Investigators discovered that many staff employed by Bangladeshi Railways were not qualified to operate a signalling system that had been recently installed. Their lack of training had caused a collision that led to nearly 600 casualties.

Right: Passengers from a second train look down on the derailed carriages of one of the two trains involved in the collision at Pubail, 15 miles (24km) west of Dhaka, the country's capital.

PURLEY, ENGLAND

MARCH 4, 1989

This fatal rear-end crash took place on a Saturday afternoon in early spring and involved a four-carriage electric train traveling from Horsham to London's Victoria main line station and a second train of eight carriages making for the same destination that had originated in Littlehampton. The manner in which the collision occurred was easily established.

Below: An aerial view of the Purley crash site. The carriages were smashed as they rolled down the steep embankment, toppling trees as they turned over.

The Littlehampton train had run through warning signals as it approached Purley and careered into the Horsham train as it was transferring from the slow to the fast track after stopping at Purley station. Six of the Littlehampton train's carriages were knocked off the track and slid down the embankment in the collision, coming to rest close by the houses that lined the foot of the slope. Some of the local residents were lucky to

escape death or serious injury from the falling wreckage but five passengers were killed and over 80 injured in the disaster.

When rail investigators attempted to uncover the causes behind the accident, they took a close look at the signalling system in the area. It was up-to-date, having been installed little more than five years prior to the smash, but had been plagued with minor problems in the recent past. These were corrected at the time but it was not inconceivable that new faults could have developed in the interim.

However, the system was examined closely by crash investigators and no obvious faults were found. It was concluded that the driver in charge of the Littlehampton train had failed to pay attention to a warning signal outside Purley and cancelled the noise emanating from his automatic warning device.

The driver in question was granted limited immunity from prosecution on the understanding that he gave evidence before the board of inquiry, but was then charged with manslaughter and endangering public life, and went on trial at London's Central Criminal Court some 18 months after the accident. He pleaded guilty to the charges brought against him, receiving a sentence of 18 months in prison, with a year suspended. However, his sentence was later reduced on appeal following a public campaign against the severity of the court's decision.

The Purley collision also highlighted some failings of the existing automatic warning system fitted to train cabs and encouraged the introduction of an automatic train protection system which would stop a train even if a driver ignored its audible danger warning.

Some trains also began to be fitted with sophisticated recording devices similar to those used in aircraft so that investigators could study the sequence of events leading up to a crash and draw conclusions from the scientific evidence generated by these "black box" recorders.

Right: One of the stricken carriages is carefully removed from the accident scene by a large crane.

Right: Police, firemen, and investigators prepare to begin their different tasks. The nearness of the wreckage to local houses clearly shows that an even severer accident was only narrowly avoided.

EXXON VALDEZ, PRINCE WILLIAM SOUND

MARCH 24, 1989

Below: A tug pulls the damaged Exxon Valdez across Prince William Sound past floating ice.

Alaska is one of the Earth's last great wildernesses – it is also rich in oil deposits. The environmental costs of any oil spillage was brought home when the *Exxon Valdez*, a 211,469-ton VLCC (Very Large Crude Carrier) ran aground on Bligh Reef as it was making its way from the terminal at Valdez down Prince William Sound to the open waters of the Gulf of Alaska.

The *Exxon Valdez* was commanded by Captain Joseph Hazelwood and had a crew of 20. Hazelwood, although young, had 10 years of seafaring experience and had made the journey from Valdez on several occasions. As was usual, he took on board a pilot, Ed Murphy, shortly before leaving Valdez at 2100 hours on March 24. Murphy was present to guide the tanker down the Valdez Narrows and past the Valdez Arm, a distance of

some 20 miles (32km). This was accomplished shortly before 2330 hours and Murphy was dropped off at Rocky Point.

After disembarking Murphy, the captain spotted a number of small icebergs (growlers) in his path and gained permission from the Vessel Traffic Control Center to alter course slightly. He then handed over control of the *Exxon Valdez* to his third mate, Greg Cousins. Cousins was asked to take the ship through a narrow seaway between Busby Island and Bligh Reef. At midnight, the ship's helmsman changed watch and

Cousins agreed to be relieved some time later. A change of course was requested at the same time but investigators later discovered that this order was not implemented for a minute.

Two minutes after midnight Cousins, checking his radar, saw that the request to change course had not been carried out and ordered a change in the angle of the ship's rudder. The helmsman had been led to believe that the new heading would be between 235 and 245 degrees, not the 247 degrees ordered by Cousins, so stopped the rudder change early. Cousins

Right: Oil from the Exxon Valdez *is pumped into another tanker in an attempt to reduce spillage.*

Right: Cleaning-up operations included the use of perforated hoses that sprayed sea water over the contaminated shorelines to wash the oil back into the sea. There it could be skimmed off with skimmers.

Below: In the port of Valdez, Alaska, workers begin the process of skimming the oil from the surface of the sea.

spotted that the tanker was dangerously close to Bligh Reef and contacted the captain. As he did so, the *Exxon Valdez* struck its rocks.

The vessel was holed and its cargo of oil began to spill into the channel. No one could guess the extent of the pollution that was to follow. Estimates indicate that over 11 million gallons (50 million liters) of oil escaped, polluting an area of 500 square miles (1300 sq km). Thick crude oil was washed ashore along over 800 miles (1300km) of Alaskan coastline.

The site of the spillage was remote and the wintry weather further complicated matters. To make things worse, one of the key anti-pollution vessels was out of action for two vital days and the air terminal building at Valdez was damaged by a storm two days after the grounding. The delays in getting the cleaning-up operation started were costly. Two weeks after the incident only around 20 percent of the spilled oil had been recovered or contained with booms and the vast oil slick was stretching up to 70 miles (115km) from the tanker. More than 30,000 seabirds as well as many mammals are thought to have died from the pollution.

Hazelwood was acquitted on charges of criminal negligence in the US but the court stated that the accident would probably not have happened if he had stayed on the *Exxon Valdez* bridge. Exxon itself was held responsible and ordered to foot the bill for the clean-up operation.

UFA, SOVIET UNION

JUNE 4, 1989

Many hundreds were killed or injured in this incident at Ufa, west of the Urals, which shocked many Russians and led to an officially sponsored day of national mourning. The incident had nothing whatsoever to do with the rail network itself but was brought about by a fracture in a natural gas pipeline lying a mile (1.6km) from the track.

Below: Wreckage, burned and twisted by the high temperatures generated by exploding natural gas, stands testament to the scale of the disaster at Ufa.

Those responsible for overseeing the running of the pipeline spotted a drop in pressure caused by the leak but rather than investigate the cause they simply increased the flow of gas through the pipeline. The vast volume of escaping gas flowed over the countryside and much of it eventually settled in low-lying ground close to the rail tracks.

As two trains approached each other on the Trans-Siberian Railway, the highly volatile gas ignited. The explosion and subsequent fireball engulfed the trains and the flames, more than a mile (1.6km) wide, moved with such speed that they threw the two locomotives off their tracks and derailed several carriages. The destruction was not confined to the trains, however. Hundreds of trees were felled and many more were stripped of their bark and leaves.

Many of those who survived the fireball suffered severe burns. Rescue services were quick to respond to the emergency and many casualties were transferred to local hospitals by helicopter. Several of the worst injured were later flown to Moscow for treatment and a number of children were sent to specialist burns units in England.

SIOUX CITY AIRPORT, IOWA, USA

JULY 19, 1989

Skilled piloting can do much to save a stricken aircraft. When the United Airlines McDonnell Douglas DC-10 crash-landed on July 19, 1989, it was only through an incredible feat of piloting that 184 of the 296 people on board the aircraft escaped with their lives.

At a height of 37,000 feet (11,277m), approximately halfway through a 930-mile (1500-km) domestic service between Denver, Colorado, and Chicago, Illinois, a loud explosion was heard from the rear of the DC-10. Instruments on the flight deck indicated that the No.2 engine mounted on the fin of the jetliner had failed. More worryingly, hydraulic pressure – which is essential to control this type of aircraft – was registering zero.

When the back-up hydraulic system also failed, the flight crew were forced to face the fact that they had a virtually uncontrollable aircraft. The only means of control available to them was to vary the thrust to the other two, wing-mounted, engines. This takes great

skill and allows little room for error. Behind the tense atmosphere on the flight deck, the passengers were instructed to prepare for a crash landing.

The aircraft was closest to Sioux City airport, where emergency services were put on full alert and all air traffic cleared from the vicinity. The aircraft pitched and rolled toward Runway 22 as the pilots fought to keep the wings level. Just 100 feet (30m) above the ground, the starboard wing dropped and struck the ground. Careering off the runway, the aircraft flipped over on its back and broke up into three pieces. It then burst into flames.

Arriving on the scene within seconds, the airport firecrews battled to free passengers hanging in their seat restraints inside, but could not prevent many from being overcome by the fumes. In all, 111 passengers and a stewardess died.

After an exhaustive investigation, it was revealed that a forged titanium alloy rotor disc had shattered. This was due to a fault in the casting process that had taken place more than 18 years previously.

Above: A United Airlines McDonnell Douglas DC-10 similar to the aircraft that crash-landed at Sioux City airport.

BOGOTA, COLOMBIA

NOVEMBER 27, 1989

Colombia is riddled with illegal drug cartels that are an agonizing thorn in the country's side. The leaders of these rackets and the killers that they employ have inflicted acts of terror on Colombian society for years. The Colombian judiciary and police agencies are the prime target, but many other innocent people have been caught in the crossfire. In 1989 a brutal act of terror was committed in the air that left the country reeling in shock, and which highlighted the need for diligent security at both national and international airports throughout the world.

On November 27, 1989, a Boeing 727-21 of Aerovias Nacionales de Colombia (AVIANCA) took off from Bogota's El Dorado airport on a scheduled flight to Cali, about 193 miles (310km) to the southwest. Among the passengers were five police informers, who between them had information that could have led to the conviction of some of Colombia's most infamous cocaine barons. They were closely guarded, and known to be under a death sentence from the cartels.

Five minutes after leaving the ground the Boeing was ripped apart by an explosion. The burning aircraft plunged to the earth and impacted on a hillside, instan-

Below: Rescue workers recovering the body of the pilot of the Avianca aircraft. The Boeing 727 crashed into a hillside shortly after take-off, killing all on board.

324

Right: The remains of the Avianca Boeing after it crashed in November 1989.

Below: Rescue workers removing the body of one of the victims from the scene of the crash.

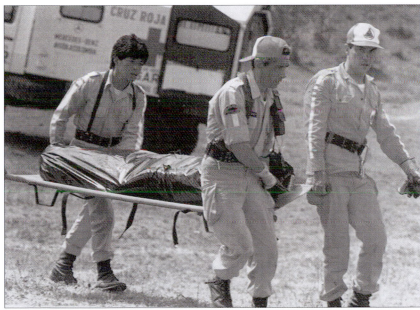

Far right: A Colombian air force specialist searches for evidence in the debris as to what may have caused the crash.

mid-point of the passenger cabin. This had ruptured the main fuel tank under the cabin floor, causing the highly inflammable aviation fuel to engulf the rear fuselage. The explosion probably also severed crucial control wires.

Police investigations have never succeeded in identifying the individual who planted the device on the aircraft. However, the senior cartel member who was widely believed to have been responsible for this act of sabotage was shot and killed by police during a raid in 1992.

taneously killing anyone who might have survived the explosion and resulting decompression. Rescue workers succeeded in recovering 110 bodies, including those of three unidentified people. These were not on the passenger list, which is supposed to provide an accurate record of all those on board. It is possible that these were the bodies of people on the ground who were killed by the aircraft.

Subsequent investigations into the crash revealed that an explosive device had detonated in a position on the right-hand side of the aircraft, somewhere near the

The recent past has shown that, despite our best efforts, disasters are an ever-present threat. Never have we been wealthier, with the money to introduce the most sophisticated technologies produced by the human mind. Yet for all this material wellbeing, catastrophes continue to hit the headlines on a regular basis. While it is true that some disasters are wholly unavoidable, many are not to a lesser or greater degree. Obviously, freak acts of nature or random terrorist attacks are difficult to predict and often impossible to prevent, but it is possible to minimize the potential risks they pose. Also, it is feasible to reduce the likelihood of major disasters brought about by technical failures, and introduce systems that can speedily and efficiently swing into action to save lives once an event has happened.

However, if a higher degree of safety is to be achieved, and it should be borne in mind that 100 percent security is an impossibility, then it boils down to two factors. First, there has to be the will to achieve the goal. This demands the constructive involvement of governments, businesses, and the wider public. Second, the enormous finances needed to achieve the ambitious aim have to be available. It remains a complex task. As the various disasters discussed in the following pages indicate, the vital consensus has yet to be reached and the money made available. For the present, our response to disasters remains a mix of partially improved safety and crisis management.

Background: *Windsor Castle, one of the British royal family's largest and most popular homes, goes up in flames on November 20, 1992.*

1990 TO THE PRESENT

SCANDINAVIAN STAR, BALTIC SEA

APRIL 7, 1990

It is rare for a ship to be the target of an arsonist. However, the *Scandinavian Star*, a 10,513-ton ferry capable of carrying up to 810 passengers, was wrecked by a deliberately-started fire. This horrifying incident made headline news around the world, and led to the tightening-up of international security and safety regulations for passenger ships.

The *Scandinavian Star* (which had several other former names) was built in Nantes, France, in 1971 and was used on a number of routes before being chartered by the Danish Da-No Line. The *Scandinavian Star* operated between Frederikshaven and Oslo.

Passengers on board at the time of the incident reported that the arsonist had struck twice. The first blaze had been spotted and extinguished. The second, however, spread rapidly and was soon out of control. Four other ferries and a number of cargo ships went to the aid of the imperilled *Star* and rescued many of the passengers and crew. Initial reports indicated that there were no fatalities, but this was not the case. Investigators discovered that at least 150 passengers and crew had died, but the figure was probably higher.

The total number of passengers aboard the *Star* will never be known as the relevant papers were destroyed in the fire and many children were probably not recorded on them anyway. The disaster raised many questions over the ferry's safety procedures: some lifeboats could not be launched and the sprinkler systems were ineffective. Because there were so many different nationalities on board – among both passengers and crew – there were communication difficulties, and in the resulting confusion people panicked and fought to get off the ship.

The fire finally burned out after four days and the still-smoldering *Star* was towed into Lysekil. The *Star* eventually made its way to Southampton and was sold to International Shipping partners in early 1994. Renamed the *Regal Voyager*, the ship was subsequently converted in Italy to carry cargo.

Below: The still-smoldering ferry, the Scandinavian Star, *in the small Swedish harbor of Lysekil, where it was towed after the disaster.*

CANTON, CHINA

OCTOBER 2, 1990

Above: The wreckage of a Chinese airliner strewn across the runway at Baiyin airport after being struck by the hijacked jetliner.

China has proved a valuable and profitable market for Boeing. This ground collision, following the hijack of one of the planes, involved three aircraft purchased from Boeing – a Xiamen Airlines Boeing Advanced 737, a China Southwest Airlines Boeing 707, and a China Southern Airlines Boeing 757.

Chinese authorities rarely divulge information on air accidents involving the dozen or so air carriers operating out of that country. Even so, China has an exemplary record of air safety, and this tragedy, the most costly in the dark history of air piracy, can be attributed only in small part to ground control procedures at Baiyin airport. The greater part of the guilt lies with the pirate himself. His desperate act cost 132 innocent people their lives.

Xiamen Airlines, one of the newer operators in China, operates regular services on a number of internal routes. A Boeing Advanced 737, one of a fleet of 737 jetliners owned by Xiamen, was making a routine domestic flight between Xiamen, in the province of Fujian, to Canton 230 miles (370km) to the southwest. At an unconfirmed time during the flight, a young man announced to cabin crew members that he had explosives strapped to his body, and demanded that the pilot reroute to Taiwan, where he presumably was seeking political asylum.

The hijacker ordered all the flight crew except the pilot out of the cockpit. The pilot attempted to reason with the young man, informing him (correctly) that the 737 had insufficient fuel reserves for the flight to Taiwan. As a compromise, the captain offered to fly to Hong Kong, but his pleas fell on deaf ears. The aircraft continued on its heading toward Canton, and then began to circle, as the captain tried in vain to convince his hijacker of the futility of his demands.

These negotiations went on for some time, until low fuel warnings began to sound on the flight deck. This left the captain with no choice but to attempt a landing at Baiyin airport, serving Canton.

Moments before touching down, the hijacker appears to have attempted to wrest control of the aircraft, causing it to thump down hard onto the runway and swerve left into a "holding area". Still traveling at considerable speed, the starboard wing of the 737 sliced into the forward fuselage of a parked China Southwest Airlines 707. The pilot, who was on the flight deck performing routine preflight checks, was only slightly injured.

Also standing in the path of the runaway jet was a China Southern Airlines 757, awaiting take-off clearance for a scheduled flight to Shanghai. Colliding with the port wing and upper central fuselage of the 757, the Xiamen 737 turned upside down and finally came to a halt. Of the 104 passengers on board the Xiamen 737, 84 died in the tragedy, along with 47 people on the 757 and the driver of a vehicle.

The Chinese authorities, who are usually reluctant to criticize their own procedures, admitted that it had been a serious mistake to allow an aircraft to taxi while a hijacked aircraft was attempting to land.

KUWAIT OIL FIELDS
FEBRUARY–NOVEMBER 1991

Above:
Firefighters
direct hoses on a
burning oil well
from behind a
fire shield.

Right: The heat
generated by the
fires was so
intense that it
was impossible
for the firefighters
to get near
without the help
of heat shields.

Early in 1991, at the end of the Gulf War, the armed forces of Iraq were forcibly driven out of Kuwait – which they had invaded during the previous year – by the combined strength of the United States and their allies. As the Iraqi forces withdrew, they loosed a telling parting shot which caused enormous damage to Kuwait's ecology and economy. On February 22, 1991, Iraqi engineers detonated thousands of pounds (kilograms) of explosives which had been laid against 810 of occupied Kuwait's 940 active oil wells. Of these, 730 exploded. Those that did not catch fire gushed crude oil uncontrollably from their shattered well heads. A total of 656 oil wells were left burning — giving a new meaning to the term "scorched earth policy."

The resulting inferno burned for nine months. The greater part of the whole country was left ablaze, and this led to the loss of up to 1.5 billion barrels of oil, worth some $27 billion. The oil flared off into the sky or became dumped in thick, tar-like lakes at the rate of about six million barrels a day. Conservative estimates put the total loss at about 67 million tons of oil. In addition, the resulting smoke plumes – which were so thick and rose so high into the air that they were clearly visible from space – produced an estimated 2.1 million tons of soot particles and more than two million tons of sulfur.

The knock-on effects were wide reaching. As far away as the Himalayas, snow became blackened with particles of soot and coal tar which had been blown right across Asia from the seat of the inferno; sheep grazing over a thousand miles (16,000km) away had their fleeces covered with a thick brown wind-borne residue. A slick of oil, immeasurably greater than that emitted from the *Exxon Valdez* in Alaska in 1989, drifted 350 miles (560km) down the Gulf coastline before it was halted at Abu Ali island. Meanwhile, in Kuwait City itself, day would be turned into night in a moment, depending on the direction of the wind. Many Kuwaitis and Iraqis were injured by inhaling lung-damaging sulfurous fumes.

And yet, the blaze, which ranks in intensity with the worse incendiary bombing raids on Dresden and Tokyo during World War II, was eventually brought under control much more quickly and painlessly than analysts had predicted. Better still, not a single fatality was added to the roll of war dead.

The first fire, in the Ahmadi Oil Field, was put out on March 18 by US Army Sergeant Forrest Irvin, who

Below: The cumbersome firefighting equipment necessary to deal with the fires was in short supply. Initially the single road from the Saudi border to Kuwait City was blocked with slow-moving, heavy civilian and military traffic. The only other way to get the equipment in was to fly it in to Kuwait international airport.

Above: A giant chimney is maneuvered over the head of a burning oil well. When the chimney was firmly bedded down over the well head, the flames coming out of the top of the chimney would be extinguished using sea water or a fire-quenching gas. The well head would then be capped.

Right: The burning oil wells belched plumes of dark, acrid smoke into the atmosphere, turning the sky dark.

tossed a grenade into the neck of the blaze and cut off the flame: the success of this technique led to it being widely used to put out the other blazes.

Firefighters from many parts of the world were rushed to Kuwait, including the US firefighting team led by the charismatic but aging Red Adair. The work of the firefighters was dirty and dangerous. Sometimes the explosives had not gone off, and had to be removed – a job that should have been done by bomb removal experts. Since there were none available the firefighters had to do it themselves. The burning wells had to be capped. Even getting near them was difficult since the heat from the blaze was terrific. Sometimes more explosives were used to extinguish the blaze.

Operation Desert Quench, as it was called, employed 1200 men, working from dawn to dusk in the desert heat. "Some wells will take three or four months to put out, and the worst are yet to come," said Red Adair in the first weeks. By July it was estimated that 80 percent of the world's oil firefighting capability was gathered in Kuwait, but even so the firefighting efforts were hampered by lack of equipment.

It was November before the last of the screaming banshee blazes was finally quelled. "It's great to hear the silence after you stop a real ripper," said Maurice Engman, a field hand with one of the firefighting companies. On November 6, the last burning well head was ceremonially extinguished by the Emir of Kuwait. But the economic damage was lasting. As a direct result of Iraq's tactics, the cost of Kuwaiti oil rose from 80 cents to between $3 and $4 per barrel.

Left: After each fire was put out, the firefighters had to cap the well to stop the crude oil gushing out. Up to 1.5 billion barrels of oil were lost – much of it being dumped on the ground – worth about $27 billion.

SHIGARAKI, JAPAN

MAY 14, 1991

Above: The badly smashed remains of one of the two trains involved in the collision at Shigaraki. Crash investigators prepare to sift through the wreckage.

Japan's railroad network has not been immune to tragedies, but they have a reputation for being safety conscious and all major incidents are treated extremely seriously. Before this fatal collision in the central region of the country, there had not been a rail disaster of such magnitude in Japan for nearly 30 years.

On November 9, 1963, the wreckage from a derailed goods train involved in a smash with a truck on a crossing was hit by two passenger trains traveling in opposite directions. The severity of this unusual incident near the city of Yokohama can be gauged from the extensive casualty list: over 160 passengers on the crowded commuter trains were killed and 120 others were injured.

The May 1991 crash, however, produced an even greater roll of casualties. A pair of diesel trains filled with commuters struck each other head-on while traveling at some speed. The impact crushed the drivers' cabs and smashed many carriages, some well beyond feasible repair. Track was torn up for a considerable distance and the trains derailed.

When rescue workers reached the scene, they were confronted by twisted wreckage and hundreds of dead and injured. As the chaos gave way to order, lists of fatalities and those injured were drawn up. The total were, respectively, 40 and 400. Although the number of dead was just a quarter of those that died in 1963, the number of injured was over three times as great, making this one of the most catastrophic events in the recent history of Japan's railroads.

SUPHAN BURI, THAILAND

MAY 26, 1991

The only accident recorded to mark the otherwise exemplary record of the Austrian Lauda-Air airline – former racing driver Niki Lauda's airline – also involved a Boeing 767, which has otherwise proved to be one of the safest commercial airliners on the market.

Quite why the thrust reverser on the port engine of the Boeing 767 deployed, as the aircraft climbed under full power through 25,000 feet (7500m), has never been established. What is clear is the catastrophic effect it had. Just 16 minutes after take-off from Bangkok airport, the airliner entered a steep dive and began to disintegrate, bursting into flames and plunging into the Thai jungle. There were no survivors among the 213 passengers and 20 crew. Wreckage was scattered over a wide area, and rescue teams found their job hampered by the density of the surrounding jungle.

A further obstacle to accurate analysis of the fatal accident was the obliteration of the digital flight data recorder – the "black box." Black boxes can sometimes hold the key to the cause of a crash; but in this case it was the cockpit voice recorder that provided the clue. When the tape from the Boeing 767 was played back, the first officer was heard to say "reverser deployed."

Working on this vital piece of evidence, investigators discovered that the most likely cause of the accident was electrical interference. The source of this is unclear, and very difficult to substantiate. It did, however, cause the thrust management to reverse the engine in flight. Reversers are only ever used to brake the aircraft on the ground run, or more obviously to reverse from a loading ramp.

As a result of this accident Boeing were forced to retrofit nearly 2000 of their jetliners with a modified thrust reversing system.

Left: Rescue workers and investigators survey some of the wreckage of the Boeing 767 that exploded and crashed in the Thai jungle.

OCEANOS, INDIAN OCEAN

AUGUST 4, 1991

Left: The last moments of the Greek liner Oceanos, *as it sinks in heavy seas off the coast of South Africa on August 4, 1991.*

The Greek-owned liner *Oceanos* had a long career during which it underwent several name changes. Originally called the *Jean Laborde*, the vessel was built in 1951 for France's Messageries Maritimes, and started in service carrying passengers and cargo between France and its African colonies. After going through several other owners, the vessel was bought by the Greek Epirotiki Lines in mid-1976 and was refitted to cater for the blossoming luxury cruise market, plying at first among the Greek islands. Some 500 passengers could be accommodated in the liner's well-appointed cabins.

The *Oceanos*'s date with destiny came during a charter cruise between East London and Durban in South Africa. On August 3, 1991, as the ship was negotiating high winds and rough seas, the captain was informed that his ship had sprung a leak in its engine room. There was a sudden power failure and it became clear that the *Oceanos* was foundering. The captain took the opportunity to abandon his ship and its passengers, supposedly to get ashore to coordinate the rescue effort. Amazingly, it was left to entertainers on board the ship to oversee the rescue. Remarkably – under the circumstances – there were no fatalities.

The evacuation, spearheaded by more than a dozen South African helicopters, was carried out smoothly and efficiently on August 4. All 580 people on board were taken off safely, the last few being removed with the aid of a Dutch container ship. The loss of the *Oceanos* – it finally slid beneath the waves bow-first in the afternoon – and the captain's behavior were a severe embarrassment to the Greek owners.

AMSTERDAM, HOLLAND

OCTOBER 4, 1992

Right: The scene in the Bijlmermeer district of Amsterdam shortly after the Boeing freighter aircraft crashed into an apartment house, partly demolishing it and setting it on fire.

Far right: A corroded connecting pin led to the loss of an engine of the Boeing 747 freighter aircraft.

Residents in the buildings adjacent to Schipol airport, serving the Dutch port of Amsterdam, have long been campaigners against the noise and smoke pollution that is for them a daily experience. In October of 1992 an accident occurred that had an even more devastating impact on their lives.

Fully-loaded, the freighter variant of Boeing's 747 can carry some 200,000 lb (90,720kg) of cargo over 5000 miles (8045km). Design tolerances include substantial safety margins to ensure that the airframe does not become overstressed. Tragically, though, it seems that the same rigorous procedures were not applied to the testing of the connecting pins securing engines and pylons on the aircraft.

El-Al Flight 1862 was loaded almost to capacity for its return flight from Amsterdam to Ben Gurion airport in Tel Aviv. As the freighter left the runway at just after 1930 hours on the evening of October 4, 1992, a damaged pin securing the No. 3 engine and pylon sheared off, causing the engine to break away from the wing pylon. At the same time the No. 4 pylon and engine were also torn off, together with part of the leading edge of the wing.

The aircraft, which was already heavily loaded, struggled to gain altitude with a substantial reduction in available power. The loss of the engine also damaged control surfaces in the wing, making the aircraft almost impossible to control. Only minutes after take-off the giant freighter smashed into the the center section of a multi-story apartment building in the Bijlmermeer district of Amsterdam.

The aircraft exploded on impact, completely destroying a large section of the building and severely damaging the surrounding structure. All four crew members and 47 of those in the apartments died in the tragedy. Hundreds more were injured.

In the investigation following the accident it transpired that the inboard midspar connecting pin that secured the No. 3 engine and pylon had corroded quite seriously, threatening the safety of the aircraft.

WINDSOR CASTLE, WINDSOR, ENGLAND

NOVEMBER 20, 1992

Art treasures beyond price were destroyed when a disastrous fire broke out in Queen Elizabeth II's private chapel in Windsor Castle where electrical rewiring work had been in progress. The blaze raged for seven hours and wrecked some of the most beautiful state apartments in the castle including St George's Hall, the scene of many magnificent banquets, which lost its fine timber roof and was gutted.

The fire was first spotted by picture restorers who were packing paintings to be stored during the latest stage of a refurbishment project which had been going on since 1989. Someone went for a fire extinguisher, but by the time he got one it was plain that the fire was serious. By now, members of the public walking past the northeast corner of the castle noticed a thin line of white smoke rising from behind the battlements. The alarm was raised at 1137 hours and at 1148 ten fire engines arrived just as the smoke turned to black – a reliable sign that the blaze had taken firm hold. As soon as the firemen in attendance realized the size of the fire, they called for reinforcements – within an hour 35 appliances had rushed to the scene from all over Berkshire and the neighboring counties of Buckinghamshire, Oxfordshire, Surrey, and London.

While firemen hosed the flames, teams of volunteers carried priceless works of art out of the inferno. Among them was Prince Andrew, Duke of York, who had been one of the first to arrive at the scene. Human

Above: The scene from Windsor Great Park, as dark smoke belches out of the castle.

One man who had been working in the dining room only a few yards from the private chapel tried to rescue some paintings, but he had to give up when he suffered burns to his hands. Later, speaking from a hospital bed, he said he had originally wanted to put out the fire but when he went to the entrance to the chapel "it was an inferno. The curtains were on fire and the walls were burning. It was all going up."

The fire leaped through the roof of the castle buildings and onlookers watched helplessly as lead in the roof melted and glass windows came crashing down. Every time the firemen got close, it flared up again and drove them back. At the height of the blaze, lorryloads of Life Guards arrived to join the rescue operation.

By the time the fire was put out, the damage was devastating. Eight valuable and historic paintings were beyond repair, all of them from the private chapel, most of the roof was destroyed and the structure of the building seriously weakened. Prince Andrew described the scene as "a very nasty mess, full of acrid smoke, warm and very wet." Although there was no loss of life and few injuries, one indirect victim of the Windsor

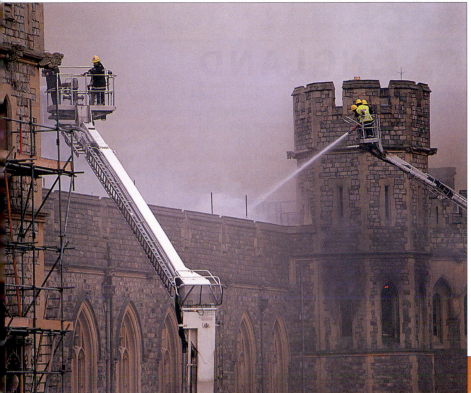

Above: Firemen tackling the blaze from high vantage points.

Right: The restoration of the state apartments of Windsor Castle cost many millions of pounds, and also gave employment to many highly skilled craftsmen.

chains passed furniture out into Engine Court and the castle courtyard soon became filled with a vast collection of ornate sofas, cabinets, and lampstands. Some items were loaded immediately into the moving vans which had come to collect the paintings, but many were left to stand for hours in the damp air.

conflagration was Christopher Lloyd, surveyor of the Queen's pictures, who was taken to hospital later that day with a suspected heart attack.

Restoration of the fire-damaged parts of Windsor Castle and 100 of its paintings was completed in late 1997 at a cost of £37 million ($60 million).

MOBILE, ALABAMA, USA

SEPTEMBER 22, 1993

This accident involved Amtrak's prestigious "Sunset Limited" and left dozens of passengers dead and many more injured. In the early hours of that fateful September day, the "Sunset Limited" crashed, was wrecked by a violent explosion, and several of its carriages plunged into a muddy creek a short distance outside Mobile. The steel bridge on which the incident occurred had been severely weakened when one of its supporting piers had been rammed by a barge in dense fog earlier in the day.

The "Sunset Limited" had entered service in April 1993 and was scheduled to make 50 halts on its trans-continental journey. The two-engined train was carrying many tourists, who were heading from Los Angeles for Miami in luxury carriages, when it approached the bridge over Cabot Bayou.

It was 0300 hours on a dark and foggy night. Because of the earlier collision involving the barge, the bridge could not support the train's weight, and distorted tracks, also caused by the barge's collision, brought about the "Sunset Limited's" derailment.

The locomotive reached the bridge at a speed of 70mph (112km/h), and the derailment and collapse led to four cars, including two passenger carriages, plunging into the deep and murky waters of the alligator-infested bayou. Others were left hanging from the shattered remains of the bridge.

Passengers struggled to make their escape through smoke-filled carriages, while others clung to the carriages that had ended up in the creek. Some were able to swim to shore or form human chains in the water to help the weaker swimmers reach the safety of the banks.

Over 40 passengers died in the accident. Many were drowned in the carriages that plunged into the bayou when its murky waters poured through smashed windows. However, thanks to the outstanding efforts

Left: An aerial view of the crash site taken some hours after the luxury "Sunset Limited" plunged into the Cabot Bayou. Rescuers had to cope with alligators.

Left: Salvage efforts are under way at the crash site. One carriage has been lifted on to a barge, while others remain where they came to rest.

of the emergency services and several brave passengers there were more than 150 survivors. The task of the rescuers was made all the more difficult by the remoteness of the crash site, the burning diesel fuel, and the poor visibility in the water. Those first on the scene had to work by the illumination provided by searchlights mounted on helicopters. It took time to get heavy equipment to the crash site because of the difficult conditions. As time passed, the rescue task became one of recovering the dead from the submerged wreckage rather than aiding the dazed and injured passengers.

Early investigations seemed to indicate that the barge's collision with the bridge had not only weakened its supports but had also distorted the track, but not sufficiently to trigger warning devices. The driver would not have had time to bring the train to a safe halt even if he had spotted the distorted track.

Left: A carriage hangs over Cabot Bayou following the crash of the "Sunset Limited," while emergency crews comb the waters for bodies and collect evidence for the investigators.

KISSIMMEE, FLORIDA, USA

NOVEMBER 26, 1993

Most railroad collisions involve two or more trains coming into violent contact because of human error or a systems failure often involving signals or some other mechanical device. However, several disasters have been brought about by a train colliding with an object, such as a rock or a piece of piping, placed on the track, either deliberately as an act of wanton vandalism or inadvertently, as in the case of a truck or tractor stalling at a crossing point.

This incident just outside Kissimmee falls into the latter category. Although there were no fatalities in this serious smash, the train in question, Amtrak's prestigious "Silver Meteor," traveling from Tampa to New York, was severely damaged in the collision and an estimated 80 of the train's 103 passengers were injured.

The late November incident began as Amtrak's "Silver Meteor" was approaching Kissimmee. Unbeknown to the train's driver, a massive electric turbine was stranded across his path and unable to advance or retreat. There was little the driver could do to avoid a collision given such a short warning of the danger ahead, and the "Silver Meteor" smashed into the turbine.

The train was derailed and its carriages suffered serious damaged. The turbine was mangled and thrown off its transporter, and a section of track was badly twisted. Emergency services responded to the incident quickly, but the wreckage strewn about took some time to clear.

Right: The remains of the "Silver Meteor" are strewn on either side of the track at Kissimmee, Florida, after its violent collision with a large generator in late November 1993.

Far right: The aftermath of a collision between Amtrak's "Silver Meteor" and a large gas turbine.

PINETOWN, SOUTH AFRICA

MARCH 8, 1994

Above: The accident at Pinetown left over 60 passengers dead and many more injured. Here, rescue workers and onlookers attend the scene of the derailment.

Accidents have been a feature on South Africa's usually safe railroad network since the earliest days of travel by train. Sometimes they have been relatively minor incidents with few serious casualties, while others have led to many deaths. As is common, these disasters have been caused by many factors, but the one at Pinetown in early 1994 involved a derailment that led to more than 60 deaths and left many passengers nursing injuries. Initial newspaper figures suggested that over 200 of those on board the commuter train needed treatment following the accident.

The commuter train came off the tracks as it was negotiating a sharp curve in the vicinity of Durban in the state of Natal. Many of its carriages keeled over into the slope that backed on to the curve as the train's forward momentum switched their weight to the outside of the curve and thus shifted their center of gravity too far over for them to return to an upright position before coming into close contact with the adjacent slope.

Although the list of casualties was awful enough, it could have been much greater if the train had been traveling on the adjacent track and the derailment had precipitated the packed carriages toppling into the fields immediately below.

South African crash and safety investigators studying the incident in great detail concluded that the derailment at Pinetown had been caused by the commuter train trying to negotiate the curved section of track at excessive speed.

ESTONIA, BALTIC SEA

SEPTEMBER 28, 1994

The *Estonia* was a large, deep-sea ferry designed for service in the Baltic Sea. It had a speed of 21 knots and could carry 2000 passengers with berths for a little over half of them. Built in Germany during the early 1980s, the vessel was originally known as the *Viking Sally* and was operated between Stockholm, Mariehamm, and Abo by Sally Lines. The ferry had two owners in the early 1990s – and two different names – before being sold on to Estonian Steamship Lines in 1992. Two years later, the *Estonia* grabbed world headlines when it went down with a horrifying loss of life. It was the worst deep-sea ferry disaster yet seen.

The Estonian Steamship Lines, a joint enterprise between the Estonian government and a Swedish company, put the *Estonia* to work on the route between Tallin, the Estonian capital, and Stockholm.

The *Estonia*'s final voyage began at 1900 hours on September 27. The ferry sailed out from Tallin but there were worrying questions over the safety of its bow doors. An inspection conducted shortly before its departure found problems with their seals, which were supposed to ensure that the doors were watertight.

Less than 90 minutes out into the Baltic, the *Estonia* encountered severe weather. Some passengers retired

Above: The passenger and car ferry Estonia, *moored in Stockholm, Sweden. Two years after being bought by Estline, the ferry went down in the Baltic with the loss of more than 850 lives.*

Left: A rescue helicopter hovers above an upturned lifeboat from the Estonia *as it searches for survivors from the ferry.*

to their cabins to sleep; others stayed in the ship's more public areas. These decisions were destined to be of crucial importance as to who would or would not survive. Around midnight an engineer on a routine inspection found water gushing in through the bow doors. The vessel's pumps were operated and began to remove the water. But so great was the volume of water entering the cargo deck that the pumps could not cope. The *Estonia* was being swamped.

At 0124 hours the *Estonia* sent out a distress signal. The waterlogged ship, buffeted by high winds and raging seas, was listing dangerously. The vessel's engines then failed. Shortly before 0200 hours the *Estonia* rolled over and sank rapidly. Those who had opted to go to bed in their cabins had no chance of escape. Those who had decided to remain in the lounge areas had a slightly better chance of survival. But there was panic on board as people fought to get off the ship.

Some people were probably crushed by heavy furnishings or by fellow passengers scrambling over

them. Some did escape but few were able to reach the safety of the lifeboats. Most of those who ended up in the icy Baltic succumbed to exposure or were drowned.

Rescue vessels knew that the *Estonia* had foundered off the coast of Turku and the first to arrive reached the spot little more than 60 minutes after receiving the distress call. It was pitch black and the seas were mountainous, yet some people were rescued. Most of these were men who probably had greater reserves of strength to cope with the cold and heavy seas. However, more than 850 passengers and crew – the final tally may never be known – were lost.

The final resting place of the *Estonia* was discovered three days later and the scene videotaped. The vessel lay in 250 feet (80m) of water and its bow door had been sheared off, probably by the violence of the storm it encountered.

A report published over three years later concluded that the ferry's bow door had a weak locking system and the heavy seas had jolted the door open.

Above: *Finnish Coast Guards carry the body of a victim ashore on the small island of Uto in the Baltic.*

ACHILLE LAURO, INDIAN OCEAN

NOVEMBER 30, 1994

Left: The Italian liner Achille Lauro *on fire 100 miles (160km) off the coast of Somalia.*

f a ship can be said to be lucky or unlucky then the *Achille Lauro* must rank as one of the unluckiest. Before being sent to the bottom by fire in 1994, the vessel had already been plagued by misfortune. In 1971 the *Achille Lauro* rammed an Italian fishing boat, leaving one of its crew dead; in 1981 two passengers died while trying to escape a fire on board the vessel; and in 1985 the ship was hijacked by Palestinians and one of its passengers, an invalid, was murdered.

The 23,629-gross ton *Achille Lauro* took nearly 10 years to build. No sooner had it been laid down for the Royal Rotterdam Lloyd Line than construction was halted by World War II. It was finally completed in 1947. The *Achille Lauro*, originally known as the *Willem Ruys*, plied the route between the Netherlands and East Indies until it transferred to sailing around the world in 1959. In the mid-1960s the ship carried migrants from Europe to Australia, but was finally bought by StarLauro of Naples for full-time cruising.

On November 30, 1994, the *Achille Lauro* was cruising off the Horn of Africa with 1000 passengers on board. A fire broke out and spread fast, forcing passengers and crew to take to the lifeboats.

The ship began to list to port as the fire continued to blaze for 48 hours. The end came as a tug was attempting to get a line on board. The ship was rocked by explosion and went to the bottom. There were reports of two casualties. Remarkably, two of the *Achille Lauro*'s sister ships – the *Lakonia* and the *Angelina Lauro* – also succumbed to fire.

Far left: Survivors from the Achille Lauro *disembark from the US cruiser* Gettysburg *at Djibouti.*

GAS PIPELINE, UKHTA, RUSSIA

APRIL 27, 1995

A fireball from a fractured Russian gas pipeline leaped thousands of feet into the air and burned for more than two hours after a spark had ignited a leak from the 55-inch (140-cm) main pipe. The flames were so fierce that some local residents fled from their homes, believing that war had broken out.

The incident happened near Ukhta, an oil-refining center 800 miles (1300km) northeast of Moscow in the semi-autonomous Komi Republic. Intense heat kept firefighters more than half a mile (0.8 km) at bay for

over two hours. The blaze forced a Japanese Airlines flight to alter course as it cruised at 31,000 feet (9300m) en route from Frankfurt to Tokyo. The pilot of the Boeing 747, Akira Yamakawa, reported: "I saw a red burning cloud below at one moment and then it became black. That was repeated again and again. We didn't know what was happening. We felt very uneasy."

Despite the ferocity of the conflagration, Russian officials said that no one had been killed or even injured. Indeed, they were at pains to play down the significance of the event – a government spokesman described it as "an ordinary fire" and added "there is no

Above: The fire from the fractured gas main was so fierce that firefighters could not get near it.

Right: The Ukhta inferno pictured at its height.

348

Right: The remains of a scorched forest in Ukhta, Russia, after a gas pipeline ignited in April, 1995, causing a blaze that ravaged the area.

need to worry." But the Tass news agency later admitted: "The Russian side routinely informed the US about the incident," presumably to allay any fears that it might have been a nuclear test.

The pipeline forms part of a 137,000-mile (219,000-km) network operated by RAO Gazprom, the Russian national gas monopoly. The lines are a constant source of concern to environmentalists all over the world because they frequently leak gas into the atmosphere. They are also liable to crack or rupture at almost any time, especially in the extreme cold of a Siberian winter. The network was built during the Soviet era and, although the pipes are made of high-grade iron imported from Germany, the PVC coating with which they are covered is only about one third of the ⅛-inch (3-mm) standard thickness to which pipelines in Europe and the US are required to conform. The welded pipeline seams are also highly vulnerable. Losses to the atmosphere from Russian pipelines are thought to exceed 10 percent every year – this is at least 10 times the level acceptable in the West.

The Ukhta explosion was almost certainly a great fire disaster, albeit one about which the full truth will probably never be known. Even if the Russian version of events is accurate, the same type of explosion might happen again, almost anywhere in the country.

Below: Smoke continued to pour from the fractured gas main in Ukhta, even after the blaze had been brought under control.

TAEGU GAS EXPLOSION, SOUTH KOREA

APRIL 28, 1995

In one of South Korea's worst peacetime disasters, at least 110 people – including 60 teenagers and 10 children – were killed by a horrific gas explosion beneath a busy road in Taegu, a city with a population of 2.2 million, 150 miles south of the capital, Seoul. More than 200 others were seriously injured.

The tragedy took place at a main intersection at about 0730 hours in the morning as heavy rush-hour traffic sat in lines waiting their turn to go through the lights. An underground railroad was being built directly beneath one of the roads at the crossing, and the normal tarmac highway had been replaced with metal sheets as a temporary surface while the excavations were carried out.

Below: At the scene of devastation following the gas explosion rescue workers search for victims.

Eyewitnesses said that when the blast came, it shook the surrounding area like an earthquake and the metal road sheets were blown high into the air by a great jet of flame. Ten houses and public buildings in the immediate vicinity of the intersection were completely destroyed in an instant; a little further away, another 60 were badly damaged.

Thirty vehicles were engulfed by flames at once, while another 30 fell over 30 feet (9m) through the temporary paving onto the underground construction site. Other buses, cars, and trucks were hurled through the air like confetti.

A vast fireball erupted from below the ground, killing rush-hour commuters and children crossing the road on their way to school. The jet of flame eventually reached more than 150 feet (45m) into the air above a 300-yard (274-m) stretch of mangled highway. The fire consumed everything in its path and, as it did so, it turned into a pall of choking black smoke.

Huge chunks of metal were thrown high into the air, and as they came back to earth over an area about a quarter of a mile (0.4km) wide, pedestrians who had not been harmed by the initial blast were killed when they were hit by lumps of debris the size of office desks. Many of the most seriously injured were pinned to the ground by metallic objects which had dropped on them from out of the sky.

One survivor, Park Yoon Ho, a 17-year-old student, heard a loud explosion on his way to school. "Then I was blown down by a gust that sent out a mushroom of dust," he said. "It looked like the explosion of an atomic bomb on television."

The disaster is thought to have been caused either when one of the tunneling drills fractured a gas pipe or when a spark from some other electrical tool ignited an existing leak.

The emergency services attempting to save the construction workers trapped underground were hampered by fractured water mains, which gushed forth sheets of water, flooding the railroad construction pit and making it difficult to work. The explosion also cut off the electricity supplies and so firefighters and ambulance men had to use their own mobile emergency generators. About 4000 soldiers, police, and firemen took part in the operation, which went on all through the following night.

Left: After the explosion many building workers were trapped in the underground railroad construction site. Here a South Korean policeman uses a rope to lift a wounded man out of the site.

SUBWAY TRAIN, BAKU, AZERBAIJAN

OCTOBER 29, 1995

Left: Narimanov underground station in Baku. A fire broke out in the tunnel between Narimanov and the next station along the line, Ulduz, during the evening rush hour.

More than 300 people died and 270 were injured, 62 of them seriously, when a crowded subway train caught fire while it was trapped in a tunnel in Baku, the capital of Azerbaijan. The main causes of death were asphyxiation and burns, but several of the victims were electrocuted as they tried to escape along the track to the nearest station.

The five-car train was packed with commuters when it juddered to a halt between Ulduz and Narimanov stations during the Saturday evening rush hour. In an instant, all the lights went out and the carriages began to fill with smoke from a fire that had already started in the tunnel. Passengers immediately began trying to escape, but the sliding doors remained firmly shut. They then attempted to break the windows, but the people were so tightly packed in the carriages that they could not get enough of a swing at the glass to break it.

Both the fire and the short circuit which stopped the train were caused by sparks from a high-voltage cable in the tunnel which had set light to the insulating material in which it was encased. This wrapping was made of PVC, which gives off acrid and highly poisonous fumes when it melts. Because it was so dangerous, insulation of this type had been removed from all

Right: *A street market in Baku, the capital of Azerbaijan. Part of the former Soviet Union, Azerbaijan retained much antiquated equipment on its underground system.*

Right: *A street market in Baku, the capital of Azerbaijan. Part of the former Soviet Union, Azerbaijan retained much antiquated equipment on its underground system.*

underground rail systems in the West by the end of the 1970s. Azerbaijan, however, was a former Soviet republic and had retained much of its antiquated equipment because it was too poor to replace it.

Soon, the fire in the tunnel set light to the middle of the stranded train. The heat in the enclosed space was so intense that it quickly melted the doors and windows and those who had been traveling in the third carriage had virtually no chance of escape.

Eventually, some of the windows did get smashed and some of the doors were prised open – survivor Gennadi Nikiferev said that when this happened "people just fell out on top of each other." But those who did get out of the train alive were still not safe – by now, the killer fumes had spread about 200 yards (183m) further along the tunnel. And there was a hidden danger – although there had been a complete power failure at the scene of the disaster, electricity was still flowing through sections of the track closer to Narimanov station – some people who did not realize this were electrocuted. Even those who did struggle to safety along the track arrived choking and vomiting and were rushed to hospital to be treated for the effects of smoke inhalation.

Survivors told horrific tales of panic and screaming and the deaths of 28 children. Manish Gurbanov, who had been in the second carriage, said: "We could not break the windows so we climbed out through a ventilation duct. I got through the tunnel by grabbing a cable on the top. People were dying all over the rails."

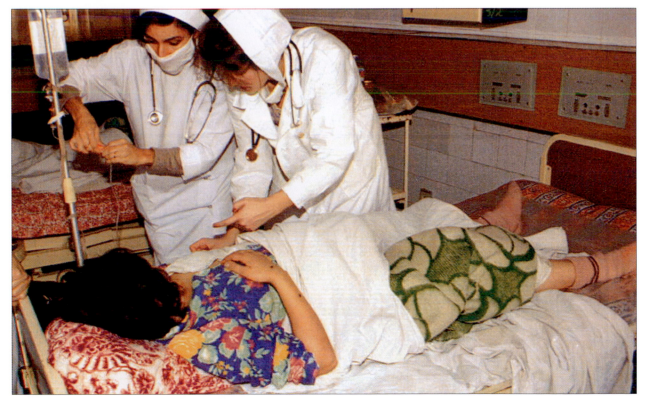

Right: *A survivor of the underground disaster is treated at the Baku Central Clinic.*

BROOKLYN SUBWAY, NEW YORK, USA

NOVEMBER 26, 1995

A New York subway clerk was badly injured when two youths lit a bottle of flammable liquid and threw it into his token booth at a station in Brooklyn, New York. The booth was destroyed and the clerk, Harry Kaufman, aged 50, suffered serious burns to 80 percent of his body.

The following day, reports of the attack appeared in newspapers and on television and radio in many parts of the world. Despite the seriousness of the crime, it would probably not have made international news had it not been for the similarities between it and a movie then on general release. *Money Train*, starring Wesley Snipes and Woody Harrelson, featured scenes – some of which were filmed in the New York subway system – in which a pyromaniac douses the insides of token booths with kerosene and then sets them alight.

Bob Dole, Republican leader of the Senate, who was then running for President of the USA, was quick to point out the link between screen violence and real-life

Above: Token booths for tickets on the New York subway. It was a token booth like this that was firebombed on November 26, 1995, badly injuring the clerk.

crime. He deplored "the pornography of violence to sell movie tickets" and said: "the American people have a right to voice their outrage . . . by derailing *Money Train* at the box office."

These views were supported by New York police commissioner William J. Bratton and Transit Authority president Alan F. Kiepper, who both said that the arson might have been inspired by the film. Meanwhile, Rudolph W. Giuliani, Mayor of New York, reacted defensively to hostile criticism of raising money by charging film makers to do location shooting. He told reporters: "The city should not be reading every script and acting as a censor."

Columbia Pictures, which produced the $60 million thriller, took a completely different view of the bombing. They issued a statement saying that they were "appalled and dismayed" by the attack. However, they denied that it was a copycat crime, claiming that the script had been based on a series of similar incidents in the early 1980s: they said that the fire had been art imitating life, rather than the other way round. As one executive put it: "We didn't invent this."

All token booths on the New York subway system are equipped with fire sensors which, when triggered either automatically or manually, emit halon from a canister to snuff out the fire and send an emergency

signal to a command center. On the night of the arson, the sensors in Mr Kaufman's booth at Kingston-Throop Avenue station had been disabled just before the attack. Some people thought that this might have been because the clerk wanted to smoke a cigarette at his desk, but the Transport Workers' Union representatives thought this sounded like an attempt to muddy the waters and smear Mr Kaufman: "I'm wondering if the Transit Authority do all the maintenance work they should on these units," said a spokesman. "Why don't we look into that?"

Far right: The entrance to a New York subway station.

Below: A scene from the film Money Train. *The film featured a scene in which an arsonist set light to a New York token booth.*

LA FENICE THEATER, VENICE, ITALY

JANUARY 29, 1996

Left: Venice's world famous opera house, La Fenice, was still smoldering the day after the blaze that substantially destroyed it. La Fenice ("The Phoenix") was no stranger to fire. The original theater had been built in the late 18th century on the site of an earlier building razed by fire, and in 1836, only 38 years after its opening, the theater was burned to the ground. It was rebuilt (hence its name, "The Phoenix") and staged the opening nights of many famous grand operas.

The Teatro la Fenice is one of the world's oldest and most famous grand opera venues. Built in the rococo style during the 18th century, it has staged the first performances of works by Verdi, Rossini, Wagner, and Stravinsky. When much of the interior was reduced to ashes late one winter's night, it was as clear a case of arson as could be imagined. Not long afterwards, the Italian authorities took the unusual step of announcing publicly that they knew how and why it happened and even, in general terms, the people responsible. Despite that, no one has ever been arrested or charged with the crime.

The fire bore the classic signs of a deliberate attack. The person or persons unknown who started it knew exactly what they were doing – they were intimately familiar with the layout of the building and laid their fire near the roof where it would cause the maximum damage in the shortest time. They struck when people with inside information would have known that the place would be unoccupied and even that the night-watchman would not be on duty. Neither, probably, was it a coincidence that the fire broke out shortly after the nearby canals had been drained for cleaning and therefore the fire barges were unable to get close enough to tackle the blaze.

The alarm was raised late at night when passers-by saw flames and smoke rising from the roof of La Fenice. Firefighting was difficult since the theater was beyond the reach of canal-borne emergency services and there are no roads in Venice. One brave helicopter pilot bombarded the fire with water which he scooped out of the Venice Lagoon.

Right: An aerial view of the theater after the main fire had been extinguished.

358

The most widely accepted and publicly aired theory was that the fire was started deliberately to avoid a contractual dispute between the owners and the builders who had been carrying out extensive refurbishment works to the interior of the theater. The contractors had fallen badly behind schedule and the destruction of a large part of La Fenice meant that the theater owners would not be able to invoke a penalty clause which, after a certain date, would have cost the builders £20,000 ($32,000) every day.

All the evidence subsequently gathered by Felice Casson, the Chief Magistrate of Venice, pointed to arson. A year later, everyone seemed to agree that it had been the work of the Mafia. But the investigation did not name names and a year later the Italian police were no nearer to making an arrest. Massimo Cacciari, the left-wing Mayor of Venice, supported Signor Casson and his findings, but complained: "Saying La Fenice was burned down by Cosa Nostra [the Mafia] is about as useful as saying that it was attacked by alien spacecraft."

Following the fire it was estimated that the new refurbishment works to La Fenice would now cost around £49 million ($80 million).

Left: *The scene of devastation inside the historic opera house after the fire.*

Right: *Fighting the blaze was made difficult by the fact that the nearby canals had been drained and it was impossible to get fire barges close to the theater.*

OFF PUERTO PLATA, DOMINICAN REPUBLIC

FEBRUARY 6, 1996

Right: US Coast Guards load debris and wreckage from the crashed Boeing 757 onto the cutter Knight Island.

Most of the 176 passengers of Flight 301 were Germans returning from a winter vacation in the Caribbean island of Puerto Plata. The Birgenair Boeing 757 had been chartered by the Dominican airline Alas Nacionales for the flight to Germany via Gander in Canada. The captain began his take-off run at about 2341 hours.

The aircraft accelerated rapidly to 92mph (148km/h) at which point the captain discovered that his ASI (airspeed indicator) was malfunctioning. The co-pilot's indicator seemed to work fine. The aircraft took off at approximately 2342 hours and began climbing to cruising altitude. As the aircraft was passing through 4700 feet (1433m) the captain's ASI was indicating 403mph (648km/h) – far in excess of the correct climb-out speed. This resulted in an autopilot/autothrottle reaction to increase the nose-up attitude and a power reduction in order to lower the airspeed.

However, the actual speed was only 253mph (408km/h). As the flight management system automatically reduced speed the crew received warnings from the audible warning system (an electronic voice which informs flight crew of potential hazards). Both pilots appear to have become confused when the co-pilot stated that although his ASI read 230mph (370km/h) they were getting an excessive speed warning. This was rapidly followed by a "stick-shaker" warning, leading both pilots to believe that both airspeed indicators were unreliable.

Finally realizing that they were losing speed and altitude, the pilots disconnected the autopilot (which had reduced the speed close to the stall speed) and applied full thrust. At 17 seconds past 2347 hours an aural ground proximity warning sounded. Eight seconds later the aircraft struck the ocean, about five miles off the coast. All those on board were killed.

In the subsequent investigation it was suggested that poor ground maintenance might have resulted in the malfunctioning of the airspeed indicator. But the crew's failure to react to the stick-shaker warning of imminent stall, and their failure to execute the procedures for recovery from the onset of loss of control, were predominantly to blame for the disaster.

SEA EMPRESS, OFF WELSH COAST

FEBRUARY 15, 1996

The Pembrokeshire coastline of South Wales is one of the most scenically outstanding in the British Isles. It is also home to a large population of grey seals and thousands of seabirds, including widgeon, shelduck, curlew, teal, and redshank. All this was to be severely threatened on the evening of February 15, 1996.

On that Thursday evening, at about 2000 hours, a 147,000-ton oil tanker called the *Sea Empress* was approaching Milford Haven estuary on the South Wales coast, carrying 128,000 tons of crude North Sea oil. The Liberian-registered tanker, with its 28-man Russian crew, was headed for the Texaco oil refinery further along the estuary. With a pilot on board, the giant 1300-foot-long (400-m) tanker was negotiating the entrance to the estuary when it hit submerged rocks and ran aground. The hull was holed, and crude oil began leaking out into the sea.

Rescue tugs and anti-pollution craft raced to the scene, and within two hours the tanker was refloated. However, a gale was blowing up, and crude oil was

Above: A bird specialist holds a rescued red throated diver covered with oil from the Sea Empress.

Above: The super-tanker Sea Empress *ran aground on submerged rocks off St Ann's Head on February 15. Despite repeated attempts to refloat the vessel, the tanker was not finally towed into Milford Haven until six days later. Meanwhile 70,000 tons of crude oil leaked from the ship.*

continuing to leak from the vessel. The crew – none of whom had been injured – remained on board the stricken tanker and attempted to pump oil out of the damaged tanks into undamaged ones.

Over the weekend the weather worsened, and during high winds on the Saturday night the towlines snapped and the *Sea Empress* ran aground for the second time. There were fears that the petroleum vapors might ignite, and the crew were taken off by RAF rescue helicopters. For the next few days salvage operations were hampered by gale-force winds. Meanwhile oil continued to leak from the tanker, spreading along the Pembrokeshire coastline, covering beaches and rocks with black slime and endangering the wildlife. Oil slicks were also moving towards the bird sanctuaries of the Skomer and Skokholm islands to the west.

A massive cleaning-up operation was mounted, with pollution experts spraying dispersant chemicals from the air over a 12-mile (19-km) oil slick. In all, about 70,000 tons of crude oil were released.

After further abortive attempts to recover the tanker, the *Sea Empress* was finally towed into Milford Haven on Wednesday, February 21, six days after it first ran aground. The vessel was still leaking oil. Six weeks later the tanker was taken to a Belfast shipyard for repairs.

Above: Tugs battle with gale-force winds in an effort to free the grounded tanker from the rocks.

Left: A Dakota plane sprays detergent over the sea to break up the oil while a helicopter hovers over the tanker prior to dropping pumps to help transfer the oil to undamaged tanks.

DUSSELDORF AIRPORT, GERMANY

APRIL 11, 1996

Left: A small fire that started in a florist's shop on the concourse of Düsseldorf airport turned into a major disaster when it melted dangerous materials that produced toxic fumes.

Sixteen people were killed and 150 injured when fire broke out in a flower shop on one of the concourses of Düsseldorf International airport. Düsseldorf is the main point of entry to the industrial Ruhr and, after Frankfurt-am-Main, its airport is the second busiest in Germany.

One of the worst aspects of this tragedy was that the fire itself was not particularly serious or life-threatening. The problem was the effect of intense heat on materials which, though flame-resistant, were nevertheless extremely dangerous because they melted and belched highly toxic gases into a confined space.

German police reported that many of the deaths were caused by inhaling the poisonous fumes emitted from molten plastic. Nine of the victims – a police officer, seven women, and a small child – were found asphyxiated in one of the elevators. The other seven bodies were found in the lavatories and in the Air France lounge – they too had all been suffocated.

The blaze started at about 1625 hours in Terminal A, the main arrivals hall for domestic and international flights of the German airline Lufthansa. Sparks were seen flying out of a ventilation grille above the florist's stall, and it is believed that these may have come from the power drill of a maintenance mechanic working above the false ceiling. The sparks quickly took hold of the plastic around the grille and became a fire which spread rapidly, sending bitter, choking smoke through ventilation shafts and passageways to many other parts of the airport.

The fire brigade was called immediately and the airport swung into its much-practiced security drill. As soon as the alarm sounded, all aircraft on the tarmac were towed as far away from the terminal as possible. The control tower radioed incoming flights with instructions to divert to Konrad Adenaeur airport, 35 miles (56km) to the north, which serves Cologne and Bonn. Passengers in the departures lounge who had been unaffected by the fire were taken immediately by coach to other airports.

Meanwhile, in the terminal building, smoke had spread rapidly through air-conditioning conduits – both the arrivals and departures halls were now enveloped in a pall of smoke – and the fire had worked its way down to the rail station beneath the terminal. Rescue teams had to wear breathing apparatus as they groped their way through the smoke-filled ruins.

Although everyone in the airport had been instructed to evacuate as soon as the alarm had been raised, there was still an unexpectedly high number of casualties. This is thought to have been because people panicked and forgot that they were supposed to head straight for the emergency exits – this explains why there were so many deaths in the elevator.

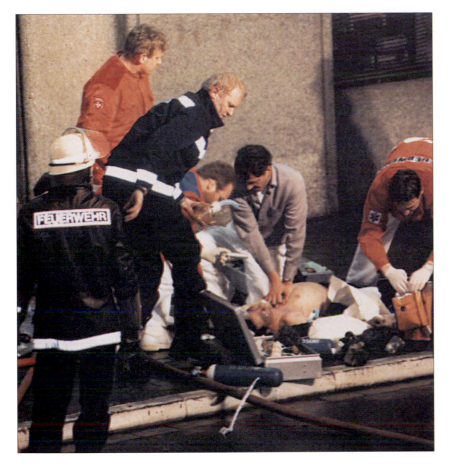

Above: *Paramedics attend to one of the injured.*

Right: *Even though the emergency services were quickly on the scene, there was a high number of casualties due to the poisonous smoke that filled the airport.*

LONG ISLAND, NEW YORK, USA

JULY 17, 1996

Above: A TWA Boeing 747 airliner shortly after take off.

Right: A wing section of the crashed 747 floating in the water on the south side of Long Island.

It took nearly 15 months to piece together the Trans World Airlines (TWA) Boeing 747 that exploded in a fireball minutes after taking off from John F. Kennedy airport in New York. The investigation by the US National Transport Safety Board was the most extensive and thorough in their history, costing in excess of $50 million, but no amount of money could ever bring back the 230 people who died on that July evening.

Eighteen months after the tragedy there remained considerable uncertainty over the cause of the accident. Early indications suggested that the giant airliner had been blown out of the sky by a bomb, or a ground-to-air missile, but as the investigation progressed this theory, for which clear evidence was lacking, receded.

Another theory, with serious ramifications for the entire Boeing 747 fleet of more than 1000 aircraft, focused on the possibility that fuel vapor in the center fuel tank (located beneath the passenger floor between the wings) ignited after it became overheated. If this was indeed what had happened, it meant that the tragedy could be repeated.

TWA Flight 800, on a scheduled service from Paris, was delayed at JFK airport for nearly three hours by mechanical problems. The center tank was virtually empty for the transatlantic flight – most of the fuel had been pumped into the wing tanks. In the sweltering July temperatures, fuel vapor would have built up above the liquid jet fuel to a temperature up to 100°F, the temperature at which the volatile fuel will ignite with the merest spark.

The aircraft finally took off and began to climb to cruising height as it flew east over Long Island. Weather conditions were near perfect. Passengers were just settling down for the long flight when, as the aircraft ascended through 13,700 feet (4100m), a huge explosion blew away a large part of the fuselage below the leading edge of the wings. The stricken 747 plunged 5000 feet (1520m) before a second explosion in the same area sealed its fate. The aircraft disintegrated, and less than 15 minutes after they had left New York, the 230 people on board were dead.

US Coast Guard, naval, and fishing vessels rushed to the scene, and for days plied the North Atlantic off Long Island, recovering the shattered remains of the aircraft. Wreckage sank into the deep offshore waters over an area of six miles (9.6km). It took until May 1997 to recover most of it.

National Transport Safety Board (NTSB) investigators were in no doubt that it was the explosion of the fuel air mixture that caused the disaster. What remained unclear was what actually triggered the explosion in the first place.

On earlier models of the 747 – the aircraft involved was nearly 25 years old – thick bundles of electrical wires run aft from the flight deck. They are thickly armored with aluminum cloth and Kevlar, and investigators found no evidence that they were split or worn. A possible answer may have been that faulty wiring in the proximity of the wing fuel tanks triggered a flame, which then entered the center fuel tank via a vent hole.

However, a theory that most senior NTSB officials supported was that jet fuel sloshing around in the center tank caused a build-up in static electricity, which then ignited the overheated fuel vapor. Like many air disasters, the search for an answer may drag on for years, but if this last theory did prove to be true, it would place many thousands of aircraft besides the Boeing 747 at risk.

Right: *A section of the fuselage is lifted from the naval salvage ship USS* Grapple *to a utility craft that will take it ashore for investigation.*

CHARKHI DADRI, NEAR NEW DELHI, INDIA

NOVEMBER 12, 1996

Above: Firemen hose down the remains of the Saudi Boeing 747 that collided in mid-air with a Kazakhstan cargo plane in November 1996.

The cargo freighter version of the Ilyushin IL-76 was developed in the late 1960s as an aircraft able to carry 40 tons of cargo to the outposts of the Soviet Union, where airstrips were often rough and ready. It entered service in 1975, and when the Soviet Union was dissolved, many former Aeroflot aircraft were transferred to Kazakhstan Airlines.

At about 1830 hours on Tuesday evening, November 12, a Saudi Boeing 747 took off from New Delhi airport

Right: On impact the two colliding aircraft disintegrated in mid-air and the wreckage was distributed over a wide area. Everyone on board the two planes was killed.

Left: Indian policemen stand guard by the debris of the Kazakhstan Ilyushin IL-76 freighter outside the village of Charkhi Dadri.

bound for Riyadh in Saudi Arabia. Many of the 289 passengers were Indian nationals returning to their jobs in Saudi Arabia. They were served by 23 crew members.

The inbound Kazakhstan Airlines aircraft, an Ilyushin IL-76, was approaching the end of its flight from the Central Asian republic of Kazakhstan. On board were 10 crew members, 27 passengers, and a considerable payload in cargo.

At 1840 hours, as the Saudi aircraft climbed through 14,000 feet (4200m) to its cruising height, the giant IL-76 collided with it. Both aircraft disintegrated almost instantaneously.

Wreckage was scattered over a wide area near the village of Charkhi Dadri, about 60 miles (96km) west of New Delhi. All those on board both aircraft – 349 people – were killed.

CHANNEL TUNNEL

NOVEMBER 18, 1996

Left: The section of the southerly Channel Tunnel with the badly damaged track and steel and concrete lining produced by the intense heat generated by the fire.

This incident could not have happened at a more inopportune moment for those in charge of the Channel Tunnel link between England and France. Although the Eurotunnel company was winning over more passengers and freight from its rivals, it was still losing some £1 million per day. Any accident could potentially shake public confidence in the link, and a closure of the system, however brief, might lead to freight returning to and staying with other forms of cross-Channel transport.

The cause of the fire, which broke out on a freight-carrying Le Shuttle traveling in the more southerly of the link's two train tunnels, remains to be discovered. However, speculation that it had been caused by an incendiary device thrown by a discontented French staff member outraged by the more than 650 redundancies announced for the link's Calais terminus seems to have been unjustified.

The normal procedure for dealing with a small fire would be for the driver to accelerate the train out of

the tunnel, but on the day in question, the fire developed so quickly that the freight train was brought to a halt within the tunnel so that its passengers could be evacuated into the central service tunnel that runs between the two main tunnels for carrying freight and passengers. This was accomplished with no little difficulty and some of those on board suffered from the inhalation of hot gases and smoke.

The rescue services had been trained to carry out their duties underground in the confines of the tunnel and eventually brought the fire under control. The locomotive itself was relatively undamaged, though badly covered with sooty deposits. However, several of the open-sided wagons used to carry large lorries – and their cargoes – were utterly destroyed. The heat of the fire in the confined space was sufficient to melt

Above: The Le Shuttle train involved in the Channel Tunnel incident lies in a siding at Coquelles, France. Its body is covered with smoke deposits from the fire.

Left: The twisted and melted remains of one of the open-sided cars used to transport goods vehicles under the Channel.

Left: The interior of the Channel Tunnel following the fire. The high temperatures were sufficient to melt the walls of the tunnel.

Below: An investigator looks over one of the open cars damaged by heat and smoke.

aluminum, and for a distance of approximately 800 yards (739m), the tunnel's concrete and steel walls suffered extreme damage. The track was also badly affected in the inferno.

It soon became clear that the impact of the fire was going to be felt for a longer period of time and cost a great deal more to repair than initial estimates suggested. Shortly after the incident, the Channel Tunnel Safety Authority in Calais took the decision to allow a severely restricted service to run along the undamaged northerly tunnel. Press reports suggested that the meeting, which lasted 14 hours, was acrimonious, but the authority finally granted Eurotunnel the right to run a maximum of six trains in each direction at 90-minute intervals once its 10-strong committee had been absolutely convinced that safety equipment,

especially fire-detection systems, was working effectively and efficiently.

Safety experts were also critical of the design of the cars used to transport freight under the Channel. The open-sided freight cars were a cheaper alternative than enclosed carriages. However, being open to the air, they could not contain a fire once it had begun and, if the train was moving at speed, the passage of wind also moving at speed would fan the flames further.

While passenger services were severely disrupted by the fire, in which it should be noted that there were no fatalities, the operators of the system could take some comfort from the fact that the public returned to using the system once it had reopened for business, and that the expected long-term drop in passenger revenue did not seemingly materialize.

COMOROS ISLAND, MADAGASCAR
NOVEMBER 23, 1996

Above: The remains of the Ethiopian Boeing 767 resting on a reef off the island of Comoros in the Indian Ocean. Rescuers found 55 survivors, including two of the terrorists, who were arrested.

The pilot of an Ethiopian Boeing 767, Captain Levl Abate, was honored by the Guild of Airline Pilots in October, 1997, for the considerable bravery and skill that he showed when his aircraft was hijacked by three terrorists.

Ethiopian Airlines Flight 961 was on a scheduled service from Ethiopia to Kenya when the hijackers burst onto the flight deck. Their demands were unclear, and the aircraft circled helplessly for a considerable length of time while the pilot tried unsuccessfully to reason with them.

The aircraft was flying in a southeasterly direction, but was soon in danger of running out of fuel. The pilot was forced to attempt a landing near Moroni, but as he approached a struggle developed on the flight deck. During the ensuing fight, both engines failed due to lack of fuel, and there was a total loss of the hydraulic power necessary to control the aircraft. At a point only some 900 feet (275m) off Le Galawa beach, and in full view of tourists (one of whom recorded the event on a cam-corder), the aircraft skimmed over a calm sea and then banked to the left, causing the port wing to touch in the wave tops. Moments later, the port engine also dragged in the water, spinning the aircraft around, and causing it to break behind the wing. Some of the passengers were thrown out by the momentum.

Onlookers and local fishermen helped shocked and dazed survivors to the shore, but 120 of the 175 people on board were killed.

PIACENZA, ITALY

JANUARY 12, 1997

This incident, which left several passengers dead and over 50 injured, involved the prestigious "Pendolino," Italy's high-speed train, as it was heading from Milan in the north of the country to the capital Rome. The "Pendolino" was scheduled to complete the journey in four hours, approximately half the time taken by regular express services. On the day of the crash, the tilting train was carrying well below its capacity, with only 150 passengers on board instead of its maximum of 900 travelers, and commentators suggested that the final casualty list could have been much higher if more people had boarded the "Pendolino" in Milan.

The Piacenza incident raised issues regarding the state railroad's intention to create a network of similar routes and trains through Italy. The ambitious plan had already been undermined to a degree when the chairman of the rail board had resigned his position after being arrested by the police the year before, charged with corruption and bribery with regard to the awarding of contracts for the proposed improvements.

The accident occurred at Piacenza, some 30 miles (48km) to the south of Milan. The train was negotiating a sharp curve as it was approaching Piacenza station and was suddenly and violently derailed. A number of the train's coaches were crushed and smashed, and

Below: The remains of some of the carriages involved in the derailment of the "Pendolino" express, January 12, 1997.

Left: One of the carriages destroyed in the crash lies across the main track at Piacenza.

some ended on their sides. Among those on board the train was Francesco Cossiga, a former Italian president. He, unlike many of his fellow passengers, emerged unscathed. Local fire and rescue services were quick to attend the scene of the crash and got to work cutting through the wreckage strewn across the track in search of trapped survivors.

The immediate cause of the derailment was a matter of speculation in the press. Some passengers stated that the train was traveling at high speed as it approached the curve when they were interviewed shortly after the crash, but there was no indication that the driver was exceeding the speed limit on the section of track where the train was derailed.

Representatives of the rail unions had a different tale to tell. They intimated that the management of the state rail network had already been warned about the possibility of inadequate safety signalling in the area of the crash and that the authorities had allegedly paid no attention to their comments. Police and security services did, however, rule out the likelihood that the accident had been brought about by Italian terrorists looking to gain publicity for their cause through an attack on a prestige target.

Right: An aerial view of the crash site gives a good indication of the ferocity of the derailment. There was considerable media speculation as to the cause of the accident.

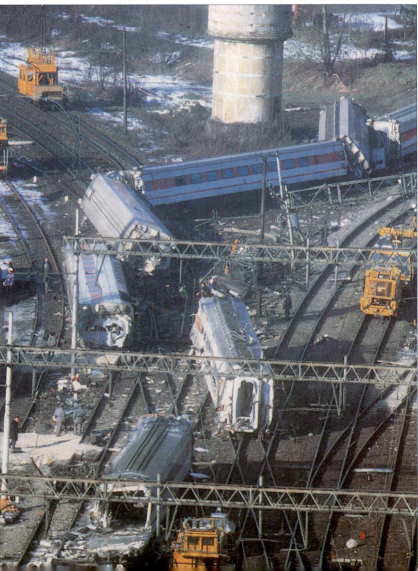

BUSH FIRES, INDONESIA

1997

For most of 1997, up to a million hectares of forest, scrubland, and plantation burned out of control across much of Sumatra and Kalimantan (the Indonesian part of the island of Borneo). This was one of the most wide-ranging fires ever and presented an enormous ongoing threat, not only to the many rare and endangered species of wildlife in its path, but also to the whole ecosystem of the planet Earth.

The logging industry is Indonesia's second largest revenue earner after oil and gas, and the government of that country is anxious to expand its palm oil, wood pulp, and rubber industries by replacing its natural forests with plantations. It plans to develop 4.4 million hectares of pulp plantations for timber by the year 2004, and 5.5 million hectares of palm oil plantations by 2000. The simplest way to clear existing forests is by setting fire to them. So these fires were started deliberately by the owners of forests and plantations – even though it has been illegal to do so since 1994.

Because of these fires, Indonesia's natural forest cover has decreased from 80 percent of the total land area in the 1960s to 57 percent today. Environmentalists fear that, if the present rate of depredation continues, the nation's forests will have been entirely laid waste by 2030.

Logged forest is more susceptible to fire than unlogged forest – this is partly because of the debris left on the ground, and partly because logging opens up the tree canopy, allowing more sunlight to enter and dry the forest floor. Moreover, much of the forest floor is made up of peat and this, combined with burning underground root systems, makes the fires much harder to control.

The smoke from these fires combined with pollutants from cities in Indonesia and Malaysia to create a suffocating smog that blocked out the sun and caused

Above:

Indonesian servicemen manhandle a hose to help a Malaysian fireman tackle the forest fires in central Sumatra. The smog resulting from the extensive fires caused severe breathing difficulties in many people in the region.

serious breathing problems to populations throughout the region. The smog contained particles of burned vegetation suspended in the air. The smoke itself was made up of numerous toxic chemicals including sulfur dioxide, hydrogen sulfide, nitrogen oxide, carbon monoxide, and ammonia.

The smog spread on the wind as far as Singapore, the Philippines, and southern Thailand and stayed in these regions even after the fires which caused them had burned themselves out: up to 70 million people are thought to have been affected.

The 1997 Indonesian forest fires were not without precedent. In 1982–1983, for example, an area of East Kalimantan about the size of Belgium (13,000 square miles/33,000 sq km) was destroyed by flames. The effects were lasting – one of the most striking is the persistent haze, which has become an annual summer feature of the weather in this part of the world.

The 1997 forest fires were made worse by El Niño, an unusual drought-inducing weather phenomenon which normally occurs once every two to seven years. Normally, the fires would have been put out by the monsoon in early autumn. In 1997 there was a severe

Above: Fire-fighters battle with the forest fires in Indonesia in October 1997.

Left: Despite the gas masks worn by these Malaysian firemen, one man had to be evacuated when the dense smog brought on an asthma attack.

Left: As the fires continue out of control, villagers in central Sumatra protect an oil pipeline that supplies southern Sumatra by throwing buckets of river water over it.

drought in southeast Asia and the fires burned on uncontrolled right through the autumn months.

This area is rich in endangered wildlife, and among the species threatened by the conflagration were the Sumatran tiger (of which there are only 400 to 500 left), the Javan rhinoceros (about 100 to 200), and the orangutan (about 30,000).

One of the greatest fears is that these fires will contribute significantly to world climate changes through the emission of massive amounts of carbon monoxide. This is a matter of grave concern to many other nations because of the greenhouse effect which reduces the oxygen supply and may shortly threaten all forms of life on Earth.

In a conference on greenhouse gas emissions in Washington, DC, in October 1997, US President Bill Clinton expressed his nation's grave concern about events in southeast Asia. "If we expect other nations to tackle the problem," he said, "then we must show them leadership." Unfortunately, there is as yet no international agreement about exactly what form this leadership should take. Japan, for example, believes that a reduction of 5 percent on 1990 emission levels by 2012 is sufficient to save the planet. By contrast, the European Union believes that target is inadequate and has called for a 15 percent reduction even sooner – no later than 2010.

Above: Indonesian soldiers and Malaysian firefighters pick up their equipment as they come to the end of a logging railroad line that has been partially destroyed by the bush fires. Many of the thousands of firefighters had to battle their way through the jungle to reach the fires they were trying to put out.

ESCHEDE, GERMANY

JUNE 2, 1998

Above: German firefighters begin the grim task of searching through the wreckage of the Munich to Hamburg express shortly after the crash at Eschede.

In the worst German rail disaster since 1947, 101 people were killed and 88 seriously injured when a high-speed train traveling at 125mph (200 km/h) hit a road bridge after being partially derailed on the approach to Eschede, some 35 miles (56km) northeast of Hannover. Wreckage containing luggage and bodies was scattered in a wide arc as carriage after carriage crumpled into the concrete supports of the bridge.

The train was the "Wilhelm Conrad Röntgen," a 10-coach, 759-seater Inter-City Express on the 0547 hours service from Munich to Hamburg, where it was due at 1300. It was carrying 350 passengers. The train left Hannover on schedule at 1033. Just after 1100, some 3.75 miles (6km) south of Eschede, passengers in the middle coaches felt an uncomfortable shaking. About a

minute later, as the front of the train emerged from beneath a bridge 1.25 miles (2km) from Eschede, the driver felt a jolt. Looking back he saw that he had lost most of his carriages. By the time he brought the power car to a halt, it was in Eschede station.

Crash investigators from the German Federal Railways Office later found part of a train wheel tire near the point at which passengers would have felt the juddering. It came from the leading axle of the rear bogie of the first coach. Although the wheel came off the rail, the train carried on for 2.5 miles (4km) until it reached a crossover 325 yards (300m) from the bridge. Here the broken wheel caught in one of the flanges in the points and derailed the coach, throwing it off to the right. This caused the next car to derail, which in turn forced Car 3 off the track and into the central support of the bridge. The train then separated between Cars 3 and 4. The bridge did not collapse completely until after the fourth car had passed under it. When it did fall, it landed on the fifth car, cutting it in half – the front section stopped about 100 yards (90m) beyond the bridge. Cars 6 and 7 were severely damaged, coming to rest almost parallel to the bridge. Cars 8 and 9 (the service and restaurant cars) were buried under the decking of the bridge, and the remainder of the train piled into the wreckage in concertina fashion. Only the rear power car avoided serious damage.

A slice of carriage measuring around eight yards (7m) hung in the air over a mass of scrap metal made up of twisted, deformed, and squashed carriages. A fireman, Walter Stroetmann, said: "This shakes me to the core. I had to collect body parts, legs and arms of men, women, and children."

Within an hour, a field hospital had been set up and the rails were swarming with more than 1000 rescue workers including trauma surgeons and border patrol personnel who helped to free passengers trapped inside the wreckage. Fifteen helicopters airlifted the injured to hospitals in Hannover. The German Army brought in heavy lifting equipment as rescuers worked into the night under arc lights. Nevertheless, the incident had taken a heavy toll in lives.

Above: German rescue staff and firefighters sift through the tangled remains of the express and bring out the bodies of the dead.

Left: Heavy lifting gear is brought in to remove the shattered remains of one of the carriages ripped apart during the fatal impact.

KHANNA, PUNJAB, INDIA

NOVEMBER 26, 1998

At least 211 people were killed and 150 seriously injured when one express rammed into the back of another that had been derailed minutes earlier on the same track.

This terrifying accident happened when the Calcutta-bound "Sealdah Express" from the garrison town of Jammu in the northern province of Kashmir collided at 0335 hours with the stationary coaches of "The Frontier Mail," which had been heading from Delhi to the Sikh holy city of Amritsar.

Part of "The Frontier Mail" had jumped the tracks near Khanna, 160 miles (256km) north of Delhi in the northern Indian state of Punjab. Sixteen of the trains' combined total of 37 coaches were wrecked in the crash that followed.

Rescue workers cut into the smashed carriages with acetylene torches to remove the dead and injured. Several hours after the collision, many of the 2500 passengers were still trapped inside the wreck, and police at the scene had to call for more cranes and other rescue equipment to deal with the worsening situation. Rail services across the Punjab were severely disrupted for more than two days while the line – one of India's busiest – was cleared and repaired.

The exact causes of the derailment and the subsequent disaster remain a mystery. It was one of many incidents, but it was the worst of nearly 300 rail accidents that occurred in 1998 on Indian railroads. The country's track network is the largest in the world and every day its 14,000 trains carry some 12 million people – not always safely.

Right: The mangled coaches of the two trains involved in the crash pictured after the end of the Indian rescue operation.

INCHON, SOUTH KOREA

OCTOBER 30, 1999

Fifty-five young people were killed and 75 injured when fire raged through a four-story shopping mall in Inchon, 30 miles (48km) west of Seoul, the capital of South Korea.

The tragedy occurred early on a Saturday evening as large groups of teenagers thronged the building's restaurants, night clubs, and an illegal beer hall to celebrate a Korean autumn festival. The blaze is thought to have been started when sparks from an electrical short circuit or a broken ceiling light fell into tins of inflammable paint thinner that had been left open in a basement karaoke bar that was being renovated at the time.

Most of the dead were found in the second-floor beer hall and in a billiard room on the third floor. The deaths were caused by smoke inhalation as panicking people rushed for a small and inadequate exit. However, safety procedures in the building were also lax in the extreme.

The building had no emergency exits, and some of the windows had been barricaded to stop people getting in without paying. The sprinkler system had been turned off. The following day police arrested an employee of the karaoke bar and four building workmen on suspicion of deactivating the basement sprinklers because they obstructed their work. They were charged with involuntary manslaughter.

The blaze was taken as further evidence that profits too often take priority over safety in South Korea. Building codes are often poorly worded and seldom enforced, and safety regulations are routinely flouted, sometimes in collusion with corrupt officials. The shopping mall blaze was the country's worst fire tragedy since 1974, when 88 people died in a burning hotel in the country's capital.

Right: South Korean policemen stand guard outside the blackened facade of the South Korean shopping mall that was ravaged by fire in October 1999.

AAMODT, NORWAY

JANUARY 4, 2000

Above: *One of the burned-out carriages of the Trondheim to Oslo express train that was involved in the incident.*

Thirty-three people were killed when two passenger trains collided head-on and burst into flames in a remote part of central Norway near the 1994 Winter Olympics town of Lillehammer. Both drivers were among the dead.

The accident happened at about 1330 hours near the village of Aamodt, some 110 miles (176km) north of the capital, Oslo. One of the trains was an express traveling south from Trondheim to Oslo. The other was a local train heading northward from Hamar. Both trains were said to have been doing about 55mph (88km/h) at the moment of impact.

Those who escaped injury tried to pull trapped passengers out of the wreckage, but their efforts were soon frustrated when leaking diesel caused one of the locomotives to explode, starting a fire that burned for

six hours. According to Ola Sonderal, an ambulance driver at the scene, some of the victims survived the crash but died in the ensuing blaze: "Inside the carriages, there were people pinned in. Many of them were conscious. But the fire that enveloped the trains left little time to get people out."

One survivor, Ben Stephenson, a British holiday-maker returning from a skiing trip, said: "We were just sitting on the train when without warning there was an almighty bang. The train just rocked onto its side and burst into flames. Inside the carriage there were those who were obviously not alive. But others were trapped and diesel was spraying all over the place. The fire was coming down the train very quickly and we needed to get people out before they were burned alive. We had to scramble out of the broken window on to the track, where we were helped by villagers."

Left: *An aerial view of the crash site at Aamodt. Heavy equipment is being brought in to begin the delicate task of removing the tangled wreckage.*

Below left: *Rescue workers search through the wreckage for the bodies of the dead on January 5, the day after the incident.*

In sub-zero temperatures, emergency services used chainsaws to clear away pine trees to ease access to the tangle of carriages. Survivors were taken to hospital in Bergen. The bodies of the victims were wrapped in green bags, loaded into hearses, and driven to Oslo for full identification.

The single track was still blocked when night fell on the 4th. Per Erik Skjefstad, of the district police, then announced: "There is no hope of finding more survivors. This is now purely a technical recovery operation." Rescue work was temporarily suspended as relatives of the victims held a memorial service and left flowers on the snow.

Crash investigators from the Norwegian National Rail Administration later said that the driver of the northbound local train had failed to stop at Rudstad station before unaccountably driving past a red light up a single track. The initial report on the incident concluded: "The accident inquiry cannot say why the train passed the signal."

It subsequently emerged that rail controllers had known that the two trains were on a collision course but could not find the right mobile phone numbers to contact the drivers and warn them. An employee of the Norwegian state railroad later said: "The telephones are moved around and used on different trains, so the lists are seldom up to date."

ENSCHEDE, NETHERLANDS
MAY 13, 2000

Twenty people died and 562 were injured when a blaze triggered a huge explosion at a fireworks storage depot in Enschede, a city in northeastern Holland near the German border. Three firemen were among the victims.

The fire started in the forecourt of a warehouse rented by SE Fireworks, a company that imported its products from China. The blaze was quite small to begin with and a number of passers-by tried to contain it themselves and then stayed to watch when the first firecrew arrived to douse the flames. But then suddenly there was a blinding flash and a massive

explosion. One survivor said: "We thought, 'We're finished.' We had no idea what was happening. All we knew was there was no place to escape. All of a sudden beer glasses shattered, window panes were blown out and people were hit by slabs of concrete."

An area of about 600 square yards (500 sq m) around the warehouse – which contained 100 tons of fireworks – was destroyed by the blast, which happened at about 1500 hours. It sent fireballs through the streets, destroying parked cars, hurling concrete fragments and glass across the city, and setting fire to several other buildings, including part of a nearby Grölsch brewery. Windows of homes 1.25 miles (2km) away

Right: An amateur video captured the moment when a huge fireball caused by exploding fireworks erupted above one of Enschede's residential areas.

Below: A lone Dutch firefighter stands amid the shattered remains of homes and cars in Enschede.

were shattered. A huge black cloud of billowing smoke plunged the immediate area into darkness and was visible 40 miles (64km) away.

Auxiliary rescue services from the Netherlands and Germany rushed to the aid of the firefighters already at the scene. Army helicopters and ambulances carried the injured to a nearby Dutch Royal Air Force base that was used as a treatment center. A local sports complex was quickly fitted out to cope with dazed and often hysterical survivors.

As police threw a cordon around the city center, where hundreds of shoppers had been showered by flying debris, firemen fought to stop the blaze from reaching gas tanks at the brewery. The fire was finally brought under control at 2130 hours.

Many of the bodies were so badly burned that they could be identified only from dental records. One rescuer found the burned body of a woman clutching two children in the cellar of what is believed to have been her home. Dutch Prime Minister Wim Kok visited the scene, which he described as "breathtakingly awful. It was as if bombs had fallen on the roofs and streets. The smell of burning is everywhere."

The day after the blaze it was revealed that the conflagration had released asbestos fumes over the whole district. Although after learning this the police refused to let any of the hundreds of evacuees return to their homes, there was criticism that the danger had not been disclosed earlier.

Investigators suspected that the blast had been caused by an arsonist. The incident was the third mysterious factory blaze in Enschede in three days. On Thursday May 11, fire gutted a plastics plant, while the following day a company producing garden furniture was also set alight.

Left: Plumes of thick acrid smoke rise from the devastated residential area of Enschede following the violent explosion.

CHILDERS, QUEENSLAND, AUSTRALIA

JUNE 22, 2000

Eighteen people died when fire tore through the Palace Hotel, a wooden two-story backpackers' hostel in Childers, some 130 miles (208km) north of Brisbane. The blaze was apparently started deliberately by a man with a grudge against the management.

Below: Police officers inspect the charred remains of the hotel's top floor verandah following the fire that devastated the hotel in Childers.

Childers, a small town with a population of 1500 people, relies on backpackers looking for work picking vegetables and other produce on nearby farms. During the season, as many as 500 backpackers from around the world arrive there every day.

Firefighters were called to the busy Palace Hotel at about 0030 hours. It took them four hours to put out the blaze. Eyewitnesses said that fire and smoke spread rapidly through the 100-year-old building and some said they heard no alarm. The hostel allegedly had only one fire extinguisher. Many of the 90 guests registered that night fled the building in their underwear as the fire blazed out of control, shattering windows. Survivors were put up in two nearby hotels and Internet cafés were cleared to allow them to send e-mails home to reassure their friends and families. The 18 bodies recovered from the ruins were taken to a forensic centre in Brisbane, where they were identified through dental records and DNA tests.

The cause of the blaze was not immediately apparent but police later learned that a backpacker had allegedly been kicked out of the Palace the week before and, as he left, supposedly told guests that they had better keep their windows open if they valued their safety. According to Rob Haddow, manager of the

Left: Floral tributes to those who died during the inferno are piled on a bench directly in front of the hotel in the immediate aftermath of the blaze.

nearby Federal Hotel: "He went berserk and the owners threw him out. He said 'I'm going to get them' and I heard he told that to people at another hotel. Then some of the backpackers said he was seen at the site last night." The police said that arson had not been ruled out but remained circumspect: "We are treating it as a crime scene but we cannot comment at this stage on possible causes," a spokeswoman said.

Then, just over a week after the fire, on July 1, a drifter was charged with attempting to murder a policeman who was arresting him on suspicion of involvement in the hostel fire. Robert Long, 37, was charged at Maryborough Hospital, some 40 miles (64 km) south of Childers, where he was treated for gunshot and dog-bite wounds. He was later transferred by helicopter to the medical wing of a prison in Brisbane.

Long, an itinerant fruit picker with a lengthy criminal history, was cornered in bushland near the township of Howard by two policemen from the Special Emergency Response Team. He allegedly lunged at one of the officers with a knife, stabbing him in the chin. He also allegedly slashed a police dog on the paw. The second officer is reported to have opened fire with a pistol, hitting Long in the arm.

Detective Inspector Jeff Oliphant, the officer heading the inquiry, said he had questioned Long about the Palace fire. Asked if Long would be charged over the blaze, Inspector Oliphant replied: "It's likely." The suspect was also charged with offences relating to fraud and theft.

Left: The remains of the staircase in the hotel that was the chief means of escape from the floor above.

PATNA, INDIA
JULY 17, 2000

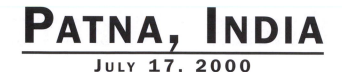

Fifty-seven passengers and several people on the ground were killed when a Boeing 737-200 crashed as the pilot attempted to make an emergency landing at Patna airport.

Alliance Air flight CD7412 flying from Calcutta to New Delhi crashed into several houses positioned just in front of the airport's runway, killing all but seven of the 58 people onboard the aircraft and at least six others on the ground.

According to eyewitnesses on the ground, one of the aircraft's engines was suddenly engulfed in a "sheet of flame" as it made its final approach to the airport. It was the Boeing pilot's second attempt at landing at Patna after he had overshot the runway a few minutes earlier. One local Patna resident who saw the final minutes of the aircraft's doomed flight announced to assembled reporters attending the crash site that "The plane was shaking and smoke was billowing from it as it hurtled down."

A mile short of the runway, the aircraft nose-dived, hit a mango grove on the airport perimeter, and then ploughed into a pair of adjoining brick houses. B.P. Bopanna, a survivor of the flight, said: "The aircraft lurched, shuddered violently, tilted to one side, and seemed to hit something with a thud." Seconds later the plane landed and he was caught in a shower of flying metal. He lay there for what seemed an eternity before being rescued.

Hundreds of local residents rushed to the scene and formed a human chain, passing buckets of water to put out the smoldering wreckage. Local emergency services searched the aircraft's remains and crash site for survivors and then began the grisly task of removing the bodies of the dead, most burned beyond recognition, from the Boeing's remains.

The aircraft, which was leased by Alliance from the state-owned Indian Airlines, was around 20 years old and was due to have been taken out of service at the end of the year.

Left: Patna residents look on as the emergency services begin the task of removing the remains of those killed in the crash from the aircraft's shattered fuselage.

FOREST FIRES, WESTERN USA

SUMMER, 2000

The year 2000 saw some of the worst outbreaks of forest fires in living memory across huge swathes of the western United States. Eleven states were ravaged, from Arizona and New Mexico in the south to Washington and Montana in the north, and from California in the west to Colorado in the east.

By early August the National Interagency Fire Center in Boise, Idaho, had received intelligence on close to 70 major fires covering some 900,000 acres (364,000 hectares). Many areas of outstanding natural beauty had suffered damage and important archeological sites, such as Colorado's Mesa Verde national park, were threatened. An ecological disaster on a virtually unprecedented scale was also in the making.

The chief cause of the inferno was "dry lightning" produced by a combination of high temperatures, low humidity, and rainless thunderstorms. Thousands of dry lightning strikes had set fire to the tinder-dry brush and undergrowth in the heavily-forested national parks and, powered by moderate winds, had spread rapidly. It was also suggested that holiday-makers vacationing deep in the forests were a contributing factor – discarded cigarette ends, carelessly watched barbecues, or burning rubbish blown by the wind were also cited as possible causes of the conflagration. In addition, there were rumors that some fires had been started deliberately. Individuals were allegedly starting random fires and then reporting them to the authorities so that they could later play an "heroic" role in the subsequent firefighting operations.

Above: Fire-retardant chemicals are dropped by helicopter on to one of the many fires that engulfed several states in the United States.

Left: A broad swathe of forestry succumbs to the flames at the height of the inferno that engulfed several western states.

Below left: Exhausted firefighters leave the scene of a blaze to take a break while their colleagues continue to battle the flames.

Whatever the causes of the inferno, the nation's firefighting machine was put under severe strain, not least because fires were springing up randomly in often barely accessible forests located in mountainous terrain. By mid-August the resources committed to stopping – or at least containing – the fires were huge. Aside from the regular firefighting teams from 46 states, units of the US Army, Air Force, and Marines had been drafted in, as had professionals from Australia, Canada, Mexico, and New Zealand. Local civilians, mainly farm workers, were also contributing to the effort. The total number of people battling the fires was estimated at more than 20,000. President Bill Clinton also released federal funds to back their efforts – something like £10 million ($15 million) per day.

The fires were tackled in a number of ways. Firebreaks were created by clearing lines in the path of the fires using chainsaws and earth movers, or by using flamethrowers in "controlled burns." Overhead, specialist firefighting helicopters and fixed-wing aircraft backed by military machines dropped flame-retardant chemicals, water, or a mixture of water and slurry at vulnerable points. Despite all of these efforts, the initial prognosis was not good. No sooner had one fire been dealt with than dry lightning strikes started others elsewhere. Experts talked of merely containing the fires rather than extinguishing them. A number were so well established it was believed they would only finally be extinguished with the onset of winter.

For all the devastation, the loss of human life was minimal. The fires mainly took place in national parks with few long-term residents, and many of them were swiftly evacuated. However, there were casualties among the firefighters. In Nevada, for example, a helicopter carrying one such team crashed, killing one man and injuring a further three.

CONCORDE, GONESSE, FRANCE

JULY 25, 2000

The ill-fated aircraft, registration code F-BTSC, was Air France's oldest Concorde, and was flying the lucrative transatlantic route. It had landed back at Charles de Gaulle airport from New York at 2130 hours the previous day and was then taken to hangar QN, where technicians worked on the aircraft's four Rolls-Royce Olympus 593 engines. The chief engineer of the Concorde had reported some strange readings in his logbook during the flight from New York. The 12-man team of technicians worked on the engines through the night and completed their task by 1230 hours on the day of departure. Take-off was planned for 1425 and at around 1400 the Concorde was towed to its boarding bay – 2B – next to the airport's control tower.

However, the Concorde did not make its scheduled departure time. Storms across Europe had delayed incoming flights and one of them, an early afternoon departure from Frankfurt, was carrying passengers who were to be transferred to Concorde. They had chartered the Concorde for a holiday of a lifetime. They were to travel to New York from where a cruise ship, the *Deutschland*, was to take them on a grand tour of the Caribbean. The Frankfurt flight touched down a little after 1500 hours and the holiday group was taken to Terminal T9 to be processed.

As the tourists arrived, the Concorde was undergoing further maintenance at the behest of its pilot, Christian Marty. Marty, 54 years old and one of only 12 pilots qualified to fly Air France's fleet of Concordes, had reported a minor fault in one of the aircraft's thrust reversers, the hydraulically operated shields that drop down over the rear of each of the four engines to slow the aircraft during landing. The problem was traced to a pump and a replacement taken from another Concorde was fitted. The switch was completed by 1515. As Marty, his co-pilot Jean Marcot, and chief engineer Gilles Jardinot completed their preflight checks, the Concorde's six-strong cabin crew boarded in preparation to greet and seat the flight's 100 passengers. By 1530 Flight AF4590 was ready for departure.

Below: A Hungarian aircraft enthusiast captures the Concorde's final moments as it attempts to gain height during its doomed take-off.

The aircraft was held at its bay for a little over 10 minutes before the control tower radioed Marty that he was cleared for take-off. At 1544 the Concorde was powering down the airport's Runway 26 when the tower spotted flames shooting from under the aircraft's port wing. The staff on duty immediately put Charles de Gaulle's emergency services on red alert, and informed Marty of what they could see. Marty undoubtedly knew of the problem – the Concorde's onboard systems would have informed him. In such situations, pilots are taught to close down the offending engine and its fuel supply. However, the aircraft was traveling at some 250mph (400km/h) and had reached V1, the speed at which it is considered safer to continue the take-off.

As the flames grew in length, Marty spoke to the tower twice in quick succession. First, he reported problems with a second engine and that the undercarriage would not retract. Second, he asked to fly to Le Bourget, some 10 miles (16km) from Charles de Gaulle. From the tower it was clear that AF4590 was in a bad way. Flying southwest, it was trailing a huge plume of smoke and flames; it was only 100 feet (30m) above the ground, and appeared unable to gain altitude.

Below Concorde lay the town of Gonesse. Overhead the aircraft began to bank to the left and lose altitude. An eyewitness saw the aircraft's nose rise and its left

Above: French firefighters douse the smoldering remains of the Concorde that crashed on the town of Gonesse.

Right: The remains of the hotel that was hit by debris from the Concorde. Several people in the hotel were killed, although others were able to make their escape.

wing slice into a stand of trees. The wing next struck the ground, flipping the aircraft on its back. It then slid along the ground and two huge explosions were reported. The center of Gonesse was unscathed by the crash, but a local hotel, the Hotelissimo, was not – four people were killed. All those on Flight AF4590 also perished, taking the total number of dead to 113.

The accident investigation began immediately and the aircraft's flight recorders were quickly recovered by site investigators. Inquiries began to focus on the aircraft's tires and fuel tanks. Evidence began to mount that two or more tires under the left wing had exploded as the aircraft powered down the runway. The theory developed that the tires, which are kept at very high pressure, had exploded, sending strips of rubber and other debris slicing into the wing and fuselage. Any debris penetrating the engine would likely lead to severe damage. Sharp blades inside the jet turbine would have been dislodged, cutting fuel lines, peppering fuel tanks, and possibly causing failure of the other engine. The combination of ruptured fuel lines and tanks coupled with hot metal would have started a fire almost immediately. Later it was announced that metal subsequently found on the runway had probably punctured the aircraft's tires.

Following the accident, Air France's and then British Airways' Concordes were grounded. It was suggested that the cost of fitting them with new safety features was far too expensive, and that the remaining aircraft would never fly again.

Left: An aerial view of the Concorde crash site taken several hours after the incident. The aircraft is shrouded in dense smoke.

BIBLIOGRAPHY

Adam, Charles Francis. *Derail: Why Trains Crash*. Channel 4 Books, 2000.

Ballard, R.D. *The Discovery of the Titanic*. Guild, Hodder & Stoughton, 1987.

Barnaby, K.C. *Some Ship Disasters and Their Causes*. A.S. Barne & Co., 1970.

Beatty, David. *The Naked Pilot: The Human Factor in Aircraft Accidents*. Airlife, 1995.

Brookes, Andrew. *Flights to Disaster*. Ian Allen, 1996.

Cahill, R.A. *Disasters at Sea: Titanic to Exxon Valdez*. Century, 1990.

Croall, James. *Fourteen Minutes: The Last Voyage of the Empress of Ireland*. Michael Joseph, 1978.

Faith, Nicholas. *Black Box: The Final Investigations*. Boxtree, 1996.

Gero, David. *Aviation Disasters*. Patrick Stevens, 1993.

Hoehing, A.A. *Great Ship Disasters*. Cowles Book Co., 1971.

Kitchenside, Geoffrey. *Great Train Disasters*. Parragon, 1997.

Launey, Andre. *Historical Air Disasters*. Ian Allen, 1967.

Schneider, Ascanio, and Arnim, Mase. *Railway Accidents of Great Britain & Europe*. David & Charles, 1968.

Schott, Ian. *World Famous Disasters*. Siena, 1996.

Semmens, Peter. *Railway Disasters of the World*. Patrick Stevens, 1994.

Stewart, Stanley. *Air Disasters*. Ian Allen, 1986.

Wade, Wyn Craig. *The Titanic: End of a Dream*. Weidenfeld & Nicolson, 1980.

ACKNOWLEDGMENTS

Adam Woolfitt/Corbis, page: 76 (top).
AFP Photo, pages: 163, 167, 189, 304 (both), 305.
AKG London, page: 109 (both).
A. Kilk, page: 315.
A. Kludas, page: 87.
Arthur Trevena, page: 51.
Associated Press, pages: 144 (both), 195, 201 (both), 206, 231, 232 (top), 278, 344.
Barry Lewis/Corbis, page: 76 (bottom).
B. Novelli, page: 314.
Boomsma Collection, page: 65.
Camera Press, page: 284 (bottom).
Caroline Penn/Corbis, page: 355 (top).
Corbis-Bettmann, pages: 5, 11, 22 (top), 35 (both), 50 (top), 69, 70, 11 (bottom), (Ryall) 114, 153, 154 (both), 290 (top), 300, 319, 312 (top).
Corbis-Bettmann/Reuters, pages: 306, 307, 334.
Corbis-Bettmann/UPI, pages: 6, 9 (top), 12, 32, 33, 45 (both), 46, 89 (top), 103 (both), 106, 107 (both), 108, 111 (top), 119, 130, 131 (both), 132, 133 (both), 143 (bottom), 151, 152 (bottom), 156 (bottom), 164, 168, 170, 173, 174 (top), 191 (top), 198, 199 (both), 208 (bottom), 217, 224 (bottom), 233, 234 (both), 237, 252, 254 (both), 294, 320, 321 (bottom).
David Muench/Corbis, page: 140.
Deutsche Presse-Agentur, page: 365 (top).
E. D. Eckstein/Corbis, page: 183.
F. O. Braynard, page: 73.
Frank Spooner Pictures, pages: 275, 272 (both), 359.
George Fryer, pages: 30, 211, 219.
Getty Images, pages: 18 (both), 19 (both), 22 (bottom), 23, 24, 36, 49 (top), 50 (bottom), 52 (both), 53, 56 (bottom), 78 (bottom), 82, 83 (top), 89 (bottom), 93, 94 (both), 97, 99, 104, 105 (both), 110, 112, 113 (both), 142, 145 (both), 156 (top), 157, 165 (both), 172, 174 (bottom), 175, 176 (right), 180, 187 (bottom), 196, 197, 204 (top), 207, 208 (top), 228 (top), 277 (top), 299 (bottom).
H. C. Casserley, page: 116 (both).
Hugh W. Cowin, pages: 177, 202, 226, 240, 262, 303, 323.

Hulton-Deutsch Collection/Corbis, pages: 8 (bottom), 28–29, 42–43, 98.
Hulton-Getty Picture Collection Ltd, pages: 7 (top), 10, 14–15, 37, 38 (both), 59, 60 (both), 66–67, 74–75, 91, 92, 117, 123 (both), 126–127, 134, 135 (both), 139, 146 (both), 148 (both), 178, 179 (both), 188, 212, 214, 215 (both), 218.
Hulton-Getty Picture Collection Ltd/Reuters, page: 343.
Joseph F. Poblete/Corbis, page: 35.
Kevin R. Morris, page: 213.
Keystone, pages: 147, 216.
Mary Evans Picture Library, pages: 16, 17 (both), 31, 34, 39, 41, 48, 49 (bottom), 54, 57 (both), 58, 68, 81, 88.
Michael S. Yamashita/Corbis, page: 194
MSI, pages: 317, 318 (both).
Museum of the City of New York/Corbis, page: 76 (top).
National Archives of America, page: 26.
Novosti (London), pages: 255, 354.
PA News, pages: 244, 245.
Photo Source/Associated Press, pages: 1, 7 (bottom).
Polfoto, page: 166.
Popperfoto, pages: 13 (both), 77, 78 (top), 83 (bottom), 85 (bottom), 96 (bottom), 102, 115 (bottom), 118, 120, 121(both), 128, 129, 136, 138, 143 (top), 144, 150 (both), 158–159, 160, 161 (both), 162 (both), 171, 176 (left), 182, 194–185, 186, 191 (bottom), 192, 193 (both), 203, 205, 225, 227, 228 (bottom), 232 (bottom), 235, 242, 243, 246, 247 (both), 261 (bottom), 263 (both), 266 (bottom), 267, 268, 274 (both), 275, 276, 277 (bottom), 281, 285, 286, 287 (bottom), 288, 289, 290 (bottom), 291, 309, 310 (both), 312 (top), 313, 324, 325 (all), 328, 329, 330 (bottom), 332 (left), 335, 342, 345 (both), 346, 347 (both), 352, 353, 355 (bottom), 358, 360 (both), 361, 362 (both), 363 (both), 366 (bottom), 367, 368 (both), 369, 377, 378 (both), 379 (both), 382, 383, 384, 385 (both), 386, 387 (both), 388, 389 (top), 390, 393, 394 (both), 395.

Popperfoto/AFP, pages: 9 (bottom), 374.
Popperfoto/Reuters, pages: 248, 280, 295, 296 (both), 301, 316 (both), 322, 333, 341 (top), 371 (both), 376 (top), 380, 381 (bottom).
Press Association, pages: 124, 125 (both).
QA Photos Ltd, page: 370, 372.
Rex Features, pages: 2–3, 64, 187 (top), 204 (bottom), 209, 210 (both), 220, 221 (both), 223, 224 (top), 229, 238 (both), 249, 250 (both), 258, 259, 260, 261 (top), 265, 266 (top), 269, 270 (top), 271, 282, 282, 284 (top), 286 (top), 292, 293, 297, 298, 299 (top), 311, 312 (bottom), 326–327, 336, 337, 338, 339 (both), 340, 341 (bottom), 348, 349, 350, 351, 364, 365 (bottom), 373, 375, 376 (bottom), 389 (bottom).
Rex Features/Sipa Press, pages:279, 308, 381 (top), 392 (both).
Rex Features/Stewart Cook, page: 391.
Ric Ergenbright/Corbis, page: 256.
Richard T. Nowitz/Corbis, page: 356.
Robert Hunt Library, pages: 25, 62, 63, 100–101, 169.
Roland Grant Archive, page: 357 (bottom).
Still Pictures, pages: 238–239.
Tony Stone Images, page: 236.
Topham Picturepoint, pages: 8 (top), 21, 61, 71, 79, 122, 155, 181 (both), 270 (bottom).
TRH Pictures, pages: 27, 40, 47, 55, 56 (top), 84, 95, 96 (top), 137, 149, 190, 241, 251, 303 (top), 366 (top).
TRH Pictures/Boeing, pages: 222, 273.
TRH Pictures/E. Nevill, page: 253.
TRH Pictures/J. Widdowson, page: 264.
TRH Pictures/Rockwell International, page: 115 (top).
TRH Pictures/US Air Force, pages: 331 (both), 332 (top).
TRH Pictures/US Army, page: 330(top).
TRH Pictures/US Department of Defense/Hughes, page: 303 (bottom).
TRH Pictures/US Navy, page: 85 (top).
Ullstein Bilderdienst, pages: 86, 90, 152 (top).
Zefa Pictures, pages: 20, 357 (bottom).